Intelligence and Artificial Intelligence

T0189787

Springer
Berlin
Heidelberg
New York
Barcelona
Budapest
Hong Kong
London
Milan
Paris
Santa Clara
Singapore
Tokyo

U. Ratsch   M. M. Richter
I.-O. Stamatescu (Eds.)

Intelligence and Artificial Intelligence

An Interdisciplinary Debate

With 18 Figures

 Springer

Dr. Ulrich Ratsch
Forschungsstätte der Evangelischen Studiengemeinschaft
Schmeilweg 5, D-69118 Heidelberg, Germany
e-mail: fest.ratsch@oln.comlink.apc.org

Professor Dr. Michael M. Richter
Fachbereich Informatik, Universität Kaiserslautern
D-67653 Kaiserslautern, Germany
e-mail: richter@informatik.uni-kl.de

Professor Dr. Ion-Olimpiu Stamatescu
Forschungsstätte der Evangelischen Studiengemeinschaft
Schmeilweg 5, D-69118 Heidelberg, Germany
and
Institut für Theoretische Physik, Universität Heidelberg
Philosophenweg 16, D-69120 Heidelberg, Germany
e-mail: stamates@thphys.uni-heidelberg.de

ISBN 978-3-642-08358-7

Springer-Verlag Berlin Heidelberg New York

Library of Congress Cataloging-in-Publication Data applied for.
Die Deutsche Bibliothek – CIP-Einheitsaufnahme.
Intelligence and artificial intelligence: an interdisciplinary debate / U. Ratsch ... (ed.). – Berlin; Heidelberg;
New York; Barcelona; Budapest; Hong Kong; London; Milan; Paris; Santa Clara; Singapore; Tokyo: Springer, 1998

Cover design: Erich Kirchner, Heidelberg

Preface

The last decade has witnessed an evolution in the field of cognition and artificial intelligence which could be characterized as an "opening up to very different points of view". The two major approaches ("paradigms") of *symbolic manipulation* and of *connectionism* no longer insist on their claims to exclusivity. They appear to have entered into a fruitful interaction, either indirectly, each becoming aware of, and trying to incorporate perspectives revealed by the other approach, or directly in concrete studies and applications. In this context there has appeared renewed interest and progress in such fields as fuzzy logics, case based reasoning, learning, etc. – but also concerning questions of meaning, language, neurobiology, and consciousness.

This situation enhances the interdisciplinary aspect which has always been an essential feature of research into cognition and artificial intelligence and raises a number of problems concerning mutual assessment and comprehension. The more researchers in each field become interested in, and motivated to learn from the methods and developments in the other fields, the better informed they must become about the latter. A further point of discussion concerns the interaction between precise theoretical developments and the rather liberal practice of applications. A renewed interdisciplinary dialog is therefore naturally emerging with an emphasis on obtaining a thorough understanding of the problems arising in the partners' field.

This book originated from a project concerning *Questions of Intelligence and Artificial Intelligence* conducted at *FESt (Forschungsstätte der Evangelischen Studiengemeinschaft – Institute for Interdisciplinary Research)*, Heidelberg. An achievement of this institution is that it provides a solid framework for research in, and the dialog between, natural sciences, social sciences, humanities, ethics, and theology and therefore can promote long term interdisciplinary research projects which would scarcely be possible elsewhere.

We highly appreciate the engagement of FESt in this indispensable activity and are thankful for its support of, and confidence in our project.

We are grateful to Wolf Beiglböck for competent advice and assistance in the completion of the book, to Angela Lahee for thoroughfull reading of the manuscript and to the other persons from the Springer staff who contributed to the publication of the book.

We want to express special thanks to Ingrid Sproll for her reliable and patient work in the preparation of the manuscript.

Last but not least, the editors would like to thank the authors of this book as well as the other participants in the above mentioned project for their highly motivated and competent contributions to the project.

Heidelberg   Ulrich Ratsch
April 1998   Michael M. Richter
                                              Ion-Olimpiu Stamatescu

Contents

Introduction

Ulrich Ratsch[1] and Ion-Olimpiu Stamatescu[1,2]

[1] Forschungsstätte der Evangelischen Studiengemeinschaft, FEST,
Schmeilweg 5, D-69118 Heidelberg, Germany.
e-mail: fest.ratsch@oln.comlink.apc.org
[2] Institut für Theoretische Physik, Universität Heidelberg,
Philosophenweg 16, D-69120 Heidelberg, Germany.
e-mail: stamates@thphys.uni-heidelberg.de

1 About This Book

Many authors have dwelt on the dispute about the future threats or promises
to be expected of Artificial Intelligence: Will AI implementations at some
time successfully mimic human intelligence and eventually even replace it?
In this book we do not intend to contribute to this debate, not only because we consider it futile. Rather, we try to bring together scientific findings
pertinent to an understanding of the mechanisms underlying "intelligent"
information processing in brains or machines. Despite the establishment of
what is called "cognitive science", investigations into the field of cognition
cannot be restricted to a single discipline. Rather, contributions from various fields of scientific endeavor are required. Results of one discipline will
more often than not complement the findings of another in a fruitful dialog.
Studies on artificial neural networks (ANNs), for instance, may yield ideas
relevant to neural activities in the brain. On the other hand, neurophysiological experiments lead to hypotheses which may be applied to ANNs. Although
results obtained with ANNs cannot directly be applied to the case of biological neural nets, they can help to deepen the insight into the performance
of brains, to increase the confidence in the plausibility of a hypothesis or to
refine it. In this way ANNs have come to be seen as tools in neurophysiological research (cf. the chapters by M. Spitzer, A. K. Engel / P. König, and
H. Horner / R. Kühn in this volume). But the possible synergetic effects of
transdisciplinary cooperation reach beyond the fields of computer science and
neurophysiology.

The contributions in this book stem from the realms of computer sciences,
psychology and neurophysiology, from mathematics and philosophy, from linguistics and physics. Out of these vast and highly diversified fields of research
only a few lines of reasoning pertinent to the understanding of "intelligence"
are presented in this introductory chapter. We shall present a few reflections
on "intelligence", some general thoughts about the techniques used by brains
or machines to achieve intelligent performance and about means to support
human intelligence by artificial tools, and a discussion of some reductionist

points of view in cognition. When referring to subsequent chapters in this book we only touch isolated aspects pertinent to the ideas presented in this introduction; we do not claim to mention the entire content of these chapters.

We first give a very brief survey of what is to follow. In her chapter "Intelligence: A Nomadic Concept" C. Karakash describes various concepts of intelligence and the fact that these concepts have constantly been changing. She also deals with the impact of artificial intelligence on the concept of intelligence and on intelligence itself. T. Spitzley's contribution on "Self-consciousness as a Philosophical Problem" presents historical positions on the notion of self-consciousness and describes the relation of self-consciousness and intelligence. The self-reference of the "I" that observes and the "I" that is observed in self-consciousness is discussed at length. F. Mühlhölzer scrutinizes Chomsky's attempt to refute Kripke's meaning skepticism. He contradicts Chomsky's "... assumption that a semantic theory should be an empirical one ...". He proposes his own "skeptical solution" to Kripke's skepticism. V. Beeh's article "Frege's Sense and Tarski's Undefinability" also deals with an uncertainty of the sense of language expressions. He discusses the relation of assignment of reference, grasping of sense, knowing truth values of terms and comprehension of the expression. He contradicts Frege's attempt to reduce the comprehension of senses to a logical analysis of truth values. In his chapter "Logics of Knowing and Believing" P. Hájek presents mathematical models that correspond to the fact that sentences of natural language are often vague. He discusses mathematical theories of "fuzzy logic" and describes the relevance of fuzzy logic to artificial intelligence. M. M. Richter's chapter "Knowledge-Based Systems" gives an introduction to various programming techniques used for knowledge-based systems, in particular describing the role of declarative programming. He also shows the way in which vague and uncertain knowledge is represented in these systems. The analysis of neural networks given by H. Horner and R. Kühn presents the theoretical background for a number of problems: architecture, learning, statistical aspects. It emphasizes features that are relevant to biological research on the functioning of brains. In particular the temporal aspects of information processing are investigated. In his contribution "Problem Solving with Neural Networks" W. Menzel explains examples of how artificial neural networks can be used to model information processing in animals (the olfactory system of the bee), the dynamics of economic processes, or composing of music. These examples are used to discuss learning procedures in ANNs. The chapter by A. K. Engel and P. König, "Paradigm Shifts in the Neurobiology of Perception", presents results of neurophysiological investigations. Focussing on visual perception, the authors also analyse the dynamic behavior of neural processing to provide a preliminary framework for the neurobiology of perception. M. Spitzer's chapter "Neural Network Models in Psychology and Psychopathology" demonstrates the use of ANN models in neuro-psychological research. Discussing the examples of language acquisition, cognitive development, hallucinations, and Alzheimer's disease, it is shown that, by comparison of experimental results and ANN models, new insights are gained which could not otherwise be obtained. The final chapter by I.-O. Stamatescu, "About Learning in Neural Networks" discusses cognitive aspects of some learning models.

2   Intelligence and Its Techniques

2.1   Features of Intelligence

Artificial intelligence (AI) is sometimes defined as the art of building machines that are able to perform tasks which, if performed by humans, would require intelligence. Plausible as this may sound, it does not tell us what "intelligence" is. There is no agreed common understanding of this notion. When trying to clarify the meaning of "artificial intelligence" as opposed to "natural intelligence", it appears to be of little relevance to have a precise definition of "intelligence". Given the spread of differing usages and the lack of consensus even about the minimal semantic elements that undeniably characterise intelligence, everyone is invited to offer his or her definition of intelligence. In default of an accepted definition we propose to adhere to a general description that leaves ample room for individual interpretations. It reads as follows: Intelligence denotes the ability of human beings to react to ever new challenges posed to them by the environment (natural or social) and to solve the problems involved.

In all sorts of activities that require intelligence the unity of perception, reasoning (in a formal, logical sense as well as in an intuitive sense), and action manifests itself. In descriptions of "intelligence" the following components are found:

Knowledge and Logical Operations in Formalized Systems A very restricted understanding of intelligence shows up in the tests of intelligence that are proposed by psychologists and educational scientists (IQ tests). In these a type of intelligence is presupposed that is measurable by a formalized set of questions. Here we shall not repeat the multitude of criticisms raised against the value of IQ tests. Suffice it to point out that these tests probe only part of the typical human capabilities. To pass the test the candidate has to prove his or her competence in performing logical operations in complex systems and the ability to systematically structure elements of knowledge. Most of the questions are of a formal nature, equivalent to solving mathematical problems. Creativity, aesthetic judgement, or rhetorical abilities, to name just a few examples, are not normally tested. Intuition may help in passing the tests, but it is by its very nature not measurable in a systematic formalized way. So intuition is not reliably checked in these tests. While intelligence certainly is required for all the tested capabilities, the totality of "intelligence" comprises more elements.

Intuition and Creativity While one may argue whether aesthetic judgement has to be considered a part of intelligence, intuition and creativity certainly have to be included. Creativity means both the ability to invent new solutions to known or unknown problems and the abilitiy to limit the number of options which have to be investigated by choosing only those that are relevant and sensible.

Social Empathy Intelligence manifests itself – some may even say arises from – social communication. Sometimes, therefore, it is claimed that empathy as well as an understanding of and the harmonious integration into social relations have to be regarded as necessary components of intelligence.

Integration of Rational and Emotional Mental Acts with Sensory and Motoric Actions It has already been pointed out that acquisition of knowledge and rational utilization of stored information are not sufficient to define intelligence. Certainly the ability to choose creatively between options is a necessary ingredient. But in addition humans make choices guided by emotions. And this is not in general to be seen as a shortcoming of the rational mind. Emotions are in many cases the guarantor for decisions which do not result in harm to the decision-maker or to other people concerned. Intelligence then lies in the wise balance of rational considerations and emotional control. Taken a bit further, the requirement is to balance sensory impressions, the control mechanisms of mental acts, as well as motoric actions and reactions. All these are channels of communication and can be utilized in a more or less intelligent way.

The series of elements sketched above not only corresponds to a more comprehensive meaning of "intelligence" than many of us are used to, it also reflects the change in understanding over the last four decades. Since the nineteen-seventies, research on artificial intelligence has contributed a good deal to changing and widening our comprehension of intelligence. In parallel to this shift, the delimitation of the realm of AI has changed, too. In the seventies and early eighties chess-playing programs, mathematical problem solvers, and expert systems were considered to belong to AI. At that time complex logical operations were seen as the ultimate accomplishments of "brains", both human or artificial. Nowadays chess-playing machines and computer programs which solve mathematical problems in an analytical form are no longer held to belong to the realm of artificial intelligence, whereas language processing, pattern recognition, and fuzzy control are major research fields of AI.

Today many computer scientists prefer not to use the term artificial intelligence. They prefer to speak about the ever more complex problems to be solved by computer software. These problems include manipulation of symbols, generating abstractions, recognising patterns, manipulating spoken inputs, and translating natural languages. Research on the last of these problems has shown, not surprisingly, that purely syntactic rules are not sufficient for transferring natural language expressions into symbols manipulated by a computer. Somehow or other the program has to learn the semantics of sequences of words and phrases. In other words, there must be a representation of the meaning of natural language sentences within the program. But this only shifts the problem from knowing what "intelligence" is to knowing what "meaning" is.

The sense of natural language expressions is frequently imprecisely determined. When, for instance, following Frege's argumentation (see the chapter by Beeh), the comprehension of the sense of expressions is reduced to looking at the truth values of their terms, many sentences of natural languages would seem to have no cognitive value. In a purely logical analysis the terms of sentences are often either contradictory or identical, i.e., they constitute tautologies. Yet, in many cases, they convey non-contradictory or non-trivial senses to a human reader; they have a cognitive value.

Uncertainty in language starts with the meaning of single terms. It is at first sight not at all clear that, when using a particular word, we mean the same today as we meant in the past. Meaning skepticism as formulated by Kripke in his interpretation of Wittgenstein relies on the fact that there are no meaning-facts to be observed like empirical facts in physics (see the chapter by Mühlhölzer.) If it was possible to lay down such meaning-facts in the protocols of observations that are reproducible in a controlled way, the semantics of words uttered now or in the past could be proved. Since, however, there is always only a finite set of empirical observations, a proof of this type cannot exist according to Kripke. Therefore, we conclude, the indeterminacy of the meaning of words and terms cannot be fixed in an empirical linguistic theory. It is not possible to prove that we attribute the same meaning to a certain word today that we attributed to it yesterday. The human brain is able to understand sentences despite the fact that their terms are not precise in the way that expressions of natural sciences are. Sentences or terms may be ambiguous or partly contradictory or tautological; their cognitive content is nevertheless understood. This obviously is not a weakness but a strength of natural languages, because it allows for the multitude of variations and nuances in the messages conveyed. Human speakers achieve a consensus about the meaning of words, be it their present or their past usage, in a conventionalist manner. The difficulties of teaching machines to understand natural languages partly stems from this fuzziness of meanings. Rule based systems (e.g. semantic nets using predicate logic to link terms) are too rigid to incorporate the diversity of expressions.

The progress of AI has not only led to a deeper insight into the differences between the operational structures of the human brain and information processing machines: In the course of these scientific findings new perspectives upon the notion of intelligence were opened. One of the seemingly unspectacular abilities of the mind is to recognize and to distinguish between objects.

A simple example may illustrate this. Children aged only a few months are able to quickly discern the faces of their parents. Computers however, which surpass every human in arithmetic calculations and most people in chess playing or deducing mathematical theorems, have great difficulties in recognizing patterns. Owing to their organizational structure – performing tasks in a series of successive steps – conventional physical symbol manipulating systems (PSS) need an extremely long time to recognize even very simple patterns. Artificial neural networks (ANN), whose organization more closely resembles the massively parallel structure of the brain, are far better suited to this task. They utilize the principle of associative memory. Obviously, the ability of associative matching of patterns belongs to the intelligence of the human mind. AI research, in particular the investigations on ANNs, has given plausible explanations of why the brain is far more efficient in recognizing patterns than conventional computers, though the single neurons are slower by orders of magnitude compared to the cycle times of computer processers. (See the contribution of H. Horner/R. Kühn)

There is another context in which to explore the meaning of this example. Looking at various manifestations of human intelligence, e.g. doing mathematics, playing chess, perceiving, recognizing and distinguishing patterns, it might be argued that there are levels of lower and higher intelligence: compared to humans computers are equally good at computation and chess, but inferior with regard to pattern recognition. Therefore, the latter might be regarded as an intellectual property of higher order. But children learn to distinguish patterns – faces of persons – long before they accomplish the skill to calculate, in contrast to this proposition. It has to be concluded, that the differences in performance between machines and human brains are not to be explained by a ranking of lower to higher intellectual acts, but by the different organizational structure.

This example illustrates how computer science may contribute insight into the relationship between the phenomenological level of intelligent behavior and the organizational structure of the basic "hardware" level of the brain. It does not necessarily imply that our understanding of intelligence or even human intelligence as such is altered by the experiences gained in AI research. But when we agree on an understanding of intelligence, which implies that intelligence together with its manifestations in reasoning, speaking etc. are formed and shaped by communication then it is not unreasonable to suggest that intelligence is changed by the reflection on itself. It is maintained in various places in this book that AI not only refers to complex performances of machines, but also is an instrument to understand aspects of human intelligence. Therefore AI in a broader sense is a manifestation of the reflection of intelligence on itself. When intelligence not only means an unalterable disposition of humans, but can be developed by communicative acts and by experiences, then the involvement in artifical intelligence may indeed change not only our understanding of intelligence but intelligence itself. C. Karakash in her contribution maintains that this is indeed the case. She suggests that human contact with artificial intelligence and traveling through "virtual realities" offer unprecedented experiences, which do not leave intelligence unchanged. New modes of intelligence emerge, according to her argument, out of the cooperation within the virtual brain of humans and machines connected globally in the internet. Thus, due to the results of AI, intelligence has to be defined in a new way, if one follows Karakash.

2.2 Comparison Between Artificial Intelligence and Human Intelligence

Let us look at similarities and differences between artificial intelligence and human intelligence (natural intelligence). This, we repeat, is not to find out if and when artificial intelligence is likely to catch up with human intelligence, but because this comparison can provide us with insights into operational principles at work in these cognitive structures.

Information Storage and Retrieval, Learning Both brains and computers can store great amounts of information. Data are recollected from computer memories at high speed and very reliably. But the precise retrieval of data presupposes that the fixed storage site is correctly addressed. If the name of a variable is miss-spelled, then the value of it will not be delivered. In brains, on the other hand, recollection of memory contents can be very unreliable, as we all have frequently experienced. Yet brains have an advantage over computer memories, since even very imprecise input data can lead to the correct retrieval of the stored information. An example is the vague description of a geographical location ("town in central Italy, up on a hill, with tall buildings like towers"), that will invoke in the mind of everyone who has visited San Gimignano a picture of that town and lead then to recall the name. To such a person "San Gimignano" appears to be the exact meaning of the description given. It has to be noted, however, that this person may err. Therefore, it is more prudent to say, that to a community of people, who have traveled through the region of Toscana, "San Gimignano" is the most probable approximation to what the speaker meant. A non-vanishing degree of uncertainty remains, there is no clear "yes-or-no" situation. Living beings are confronted with changing environments and fuzzy signals, and they have to react in a purposeful adequate manner. It is an evolutionary advantage to associate exact meanings with imprecise input information and to recognize similar but not identical situations as equivalent. It is advantageous for the cat to recognize a mouse of which only the tail is visible. It is of vital importance to the prey to recognize a predator of which only the ears and eyes are visible. The world of experience of these animals is relatively simple. The section of it relevant to the examples consists of predators and of prey. Here yes-or-no decisions are vital and therefore adequate. The human world of experience is far more complicated; it consists of many situations, where yes-or-no decisions are neither necessary nor adequate. It may not be of vital importance if the person who depicted an image of an Italian town with high towers really had "San Gimignano" in mind. It may, however, be vital that the listener can easily accept that "San Gimignano" was in fact not meant.

With regard to the principles underlying the method of storage and retrieval, there is a significant difference between symbol manipulating machines and artificial neural networks. In conventional data processing machines (PSS), the single data elements are stored in precisely located addresses of the memory. Pieces of information are retrieved by referring to this address, regardless of the particular type of information. In artificial neural networks, however, stored information is spread over the set of synaptic weights in the entire net. Storage of information is performed in a process of learning. A group of "input neurons" of the net are presented with a pattern ("input vector") which is an isomorphic map of the information to be stored. As a consequence a great number of synaptic weights are changed. When subsequently the input neurons are presented with a closely similar pattern, not necessarily an identical one, the net matches the activity of the neurons in an associative way, so that the "output neurons" exhibit a pattern which represents the information to be retrieved. The information to be stored or to be retrieved is, so to speak, itself the address of the memory. See the articles of W. Menzel, of H. Horner and R. Kühn, of M. Spitzer, and of I.-O. Stamatescu for details.

This is very similar to the way in which the brain stores and retrieves information. In machines learning has to be stimulated from the outside by a human trainer. The brains of living beings receive the stimuli for learning and knowledge acquisition through their sensory systems. The continuous contact with the outside world delivers to the brain a constant stream of patterns to be stored in the brain and to be compared with already stored pieces of knowledge. Learning is equivalent to a reorganization of synaptic weights. In biological brains they are changed in a self-organizing process. In artificial neural networks the same effect is achieved by special procedures (e.g. error back-propagation) that are programmed by the human user (see again the articles of W. Menzel, of H. Horner and R. Kühn, of M. Spitzer, and of I.-O. Stamatescu).

In a sense, allowing for a certain amount of imprecision turns out to be an advantage. Looking at the learning procedures of ANNs, it is found that in many cases it is useful to refrain from the highest possible degree of accuracy and from the highest possible velocity in learning steps. When training ANNs, the adaptation of synaptic weights is deliberately restricted to small steps, even in cases where adaptation to the desired output state could be achieved in a single step. There are various reasons for this. Here we want to mention only two of them. Consider a net which is supposed to store a number of patterns. The patterns are successively presented to the input layer of the net. If the change of synaptic weights in each step were to lead to the optimal adaptation to the presented pattern, then only this pattern could be retrieved. Letting the net adapt only a small step in the direction of the optimal state, guarantees that after a number of cycles (presenting all patterns in each cycle) the net "remembers" all the patterns. The synaptic weights achieved by this process are not identical to those optimal for any one of the patterns. It is the deviation from the individual optimal states that ensures the ability of the net to store many patterns.

Also, when the set of patterns consists of groups of similar patterns, learning by small steps allows the net to "recognize" the similarities. Thus it is able to recognize patterns it has not been presented before by their similarity to a group of patterns which were presented during the training process. Again, the fuzziness in the adaptation to single patterns and also the slowness of the adaptation turn out to be of benefit. (For a discussion see the chapters by M. Spitzer and I.-O. Stamatescu.)

Interaction with the Outside World, Knowledge Acquisition For information to become knowledge, a minimum requirement is that the single chunk of information be set in a context of related information ("semantic network"). In well-ordered referential databases, the storage of information is organized in this way. Both in physical symbol manipulating machines and in artificial neural networks, it is the task of the programmer to ensure this type of organization. An essential feature of human knowledge is that it does not only rely upon a database of pure facts. Instead, the facts stored in the memory are linked to opinions, intentions, normative valuations, and emotions. These cognitive connections are a result of experience gained in the social and natural environment. This comprehensive knowledge base is continuously refined due to the further interaction with that environment. In machines, so far, only little of this capacity could be implemented.

Machines have no continuous interaction with the outside world and, more important, the interaction is passive, and not controlled by the machhine itself, whereas living brains actively seek information to be processed. Conventional computers are fed with the data that the respective algorithm needs, deliver the results of the algorithmic operation, and stop. ANNs, too, are given the input data and, after rearranging the synaptic connections, exhibit some output pattern and stop. Normally neither of these machines uses the results of one operational cycle to start a self-defined new task. For biological brains, however, there are no clearly defined starting or terminating points. One reason for the brain's continuous activity is the permanent exchange with the environment by way of sensory input and motoric actions/reactions. Through these the brain obtains ever new incentives for actions, often more at a time, which are not performed sequentially, but in parallel.

Self-consciousness A further reason why the brain is active in an uninterrupted way comes from the inside. Some would say from the human mind.

Others contend that the mind is but an epiphenomenon of the brain and would therefore say: the impulse stems from the brain itself. Human intelligence is inextricably linked to self-consciousness, e.g. to the human capacity to reflect on ones own thinking, to be aware of each mental act. Machines as we know them certainly do not possess this ability. The article of T. Spitzley gives an overview of the philosophical notion of self-consciousness. But perhaps it is justified to ask for traces of "self-consciousness" in computers that are supposed to solve problems in an "intelligent" way. Self-consciousness in human minds is the quality that allows self-control. Being aware of my thoughts is a necessary prerequisite for controlling my line of thinking, choosing amongst alternatives, suppressing particular paths of thinking, switching ideas, in short: creatively controlling mental acts.

Most computer programs follow predetermined rules and instructions. Even switching between alternative procedures and branching of program paths including the conditions under which certain alternatives are chosen, are fixed by the programmer in advance. So far only few programs exist which attempt to change their own structure while running. It remains to be seen whether artificial neural networks open new opportunities for computers to gain more flexible "self-control". So, up to now, mental acts of human brains still differ from procedures of information processing in computers in at least one decisive way: The human being is aware of the mental act – be it thinking, feeling, or any other mental function –, whereas the computer is not. Following an argumentation similar to the one given in the previous paragraph it is often argued that self-consciousness is a prerequisite for intelligence or even a constitutive factor of the mind's ability to appoint sense or meaning to a sequence of symbols, e.g. words of a sentence.

We have already mentioned the two main paradigms of AI: the physical symbol-manipulating systems (PSS) and the artificial neural networks (ANN). There are many applications belonging to at least one of them: inference machines, expert systems, language processing programs, in particular grammar formalisms (parsing algorithms) and grammar learning ANNs, image recognition and processing systems, associative memories, fuzzy controllers, speech perception ANNs and others. In the previous sections we have repeatedly pointed to the role of "fuzziness" in mental acts. For learning in brains and ANNs it is viewed as important that the neural net does not exactly adapt to the input pattern; a constitutive feature of natural languages are imprecise terms and uncertain reference of sentences. Many sentences do not have a truth value of either 0 or 1. For an expression like "the tree is tall" the "truth" lies rather somewhere in between. Two-valued logic is not adequate for a predicate calculus to be applied to statements like this, i.e., to a significant part of all statements in natural languages. Reflecting this and similar situations, the mathematical theories of many-valued logic ("fuzzy logic") are being applied in AI systems. The contribution of P. Hájek describes mathematical formalisms of fuzzy logic as the "Logics of knowing and believing", as well as their application in AI.

Conventional rule based PSSs have been too rigid to suit the plasticity of languages and other manifestations of intelligence. Realizing the importance of "fuzziness", ways to overcome this drawback are being investigated. In his article M. Richter shows how knowledge-based systems are programmed to cope with vague and uncertain knowledge. Fuzzy logic plays a major role together with "declarative programmimg". ANNs by their very nature are better suited to represent uncertain concepts. That is one of the reasons why ANNs have attracted renewed attention and have been intensively investigated since the early eighties. At that time there was a dispute about which paradigm was the best, PSS or ANN. Today we see a convergence of these. To a certain extent rules have to be implemented in the construction of neural nets which are intended to model, for instance, the dynamics of the stock exchange (see the article of Menzel). On the other hand, as just mentioned, fuzziness is introduced into knowledge-based systems and in expert systems qualitative reasoning, too, is used. In fact the question is no longer "which system is better?", nor is there merely a sort of tolerance and coexistence, letting each system occupy its field of relevance. Rather, we see systems emerging that take advantage of both paradigms by combining them in a complementary way. Karakash contends: "Although they [the two paradigms] are antinomic, the two approaches are probably complementary". Their further development may show them to be not even antinomic. In the following we shall also touch this point.

3   Reductionist Points of View in Cognition and the Neural Network Approach

3.1   Paradigms and Reductionist Hypotheses

For a long time the concepts of "artificial intelligence" and "cognitive psychology" have been intimately connected to what is known as the "symbol-manipulation paradigm". The "connectionist paradigm" on the other hand is typically associated with neuroscience and neurophilosophy, marking the difference of perspective in approaching similar questions of cognition and behavior. Nevertheless both the symbol-manipulation paradigm and the connectionist paradigm represent equally justified, well defined reductionist hypotheses in cognitive science – and corresponding approaches to the question of machine "intelligence" – and both must be questioned concerning their validity and usefulness.

A way to understand the split might be to project back the two developments to the point represented by the seminal paper of McCulloch and Pitts (1943) "A logical calculus of the ideas immanent in nervous activity", where in some sense both paradigms can be found. As is well known, the paper shows that starting from idealized neurons modeled as two-state units (active/inactive) and connected by excitatory or inhibitory "synapses" one can reproduce arbitrarily complex logical functions. Thereby activity represents "true" and rest represents "false" and a synapse contributes to excite or inhibit the postsynaptic neuron if the presynaptic neuron is active and has no influence if the latter is at rest. A neuron fires if the sum of afferent synaptic contributions exceeds a given threshold.

The reduction suggested by this analysis is contained in the authors' remark: ".. Thus both the formal and the final aspects of that activity which we are wont to call *mental* are rigorously deducible from present neurophysiology." This can be further specified in one of two ways: one can conjecture that the logic function is the fundament of thinking (directly relying on the brain's activity) and then demanding that this function be reproduced on any substrate – the "physical symbol system – hypothesis", a "top-down" approach. Alternatively, one can propose to reproduce thinking from highly interconnected assemblies of simple "cells" (regarded as simplified brain models) – the "connectionist hypothesis", which is a "bottom-up" approach.

Now, a reductionist procedure tries to identify structural principles relating the description of the complex phenomena under discussion to some "more fundamental" description. This latter might be defined in a pragmatic manner as the most elementary level which still allows one to identify how the more complex phenomena can be deduced starting from it. In the present case the "complex phenomena under discussion" are (human) thinking and the reproduction of (some of) its features in artificial systems. In the case of the physical symbol system hypothesis the more fundamental level is symbol manipulation, i.e., logic. For the connectionist hypothesis the more fundamental level is interaction and connectivity, i.e., physics. Both these positions must acknowledge deep limitations in their reductionist scope.

The "physical symbol system" paradigm meets an essential difficulty in treating context, common sense reasoning, natural languages, induction, and associative recognition, etc., which are notoriously refractory to logical formalization. However these may be only symptoms of a deeper inadequacy. The main arguments against the "symbol manipulation" reduction can be tentatively classified as follows:

1. *Semantic*: symbol manipulation does not ensure "understanding" since meaning cannot be reduced to the relations between symbols. In fact symbols are supposed to have to each other relations compatible to those between the observations concerning the things they symbolize[1]. But nei-

---

[1] As expressed by H. Hertz a century ago. So for instance the appearence of the electric charge and of the electromagnetic field in the equations of electrodynamics put them into a relation which can be tested experimentally. And we also know that a relation between symbols obtained by mathematical reasoning can point to a relation between things before it is found experimentally, like the existence of the antiparticles.

ther does such a system – unless it is very simple – seem free of contradictions and closed, nor does it seem possible to stabilize it without reference to empirical knowledge. The development of physical theories is particularly instructive in this sense, especially because of the important role played by mathematics.

2. *Cognitive*: induction is just one example of an aspect which does not fit into a conventional logical scheme. It may be argued that the capabilities of the human intellect involve "non-algorithmic" procedures, which cannot be solved logically. They can be observed in mathematical thinking, e.g. Cantor's "Diagonalverfahren" and similar procedures connected to Turing or Goedel analysis – but in fact they also underlie much of our day-to-day thinking.

3. *Biological*: intellectual activity cannot be separated from its "substrate": in the more trivial sense this means the body, in the more refined it involves also evolution and history. The generation of motivation and intentionality may also be involved here.

The "connectionist" approach on the other hand has an essential difficulty in taking into account concepts, strategies and, generally, any high level structure. Again, the criticisms can be divided in various classes:

1. *Explanative*: a connectionist simulation does not itself provide understanding, it must be supplemented by a high-level analysis in terms of tasks, rules, concepts, – i.e., in the end, an analysis at the level of symbolic structures. But this is difficult as long as there is no clear way of submitting a connectionist system to such a description.

2. *Communicative*: there is no "intersubjective" communication, in particular exchange of "strategies", "achievements", "understanding" or any other efficient, structured information between connectionist systems – the most one can do is to take over the architecture and connectivity matrix. This means at the same time that it is difficult to build in information obtained in any other way.

3. *Constructive*: any connectionist system must begin with an abstraction about the architecture. But it is difficult to be sure that the relevant structure of brain functioning has been grasped in this abstraction. Since this "connection" is the main argument for the adequacy of connectionist models for cognitive problems, one can never be sure that the processes simulated in this way do not miss essential features needed for simulating "thinking" in any way.

4. *Performative*: without a symbolic description the claim that a connectionist system realizes thought processes cannot be tested, one can at most envisage a Turing test – with its problematic character.

Both hypotheses are very vague about dealing with "intentionality", "motivation", etc. But the criticism from this direction also has an essential difficulty in defining what it means by these terms – unless it claims that "folk psychology" should have the status of a theory, which is a very disputable move (see, e.g. the criticism by Churchland 1989).

Finally, the original (McCulloch and Pitts' 1943) position is in a sense less formal, since it relates two levels open to empirical investigation. But it can be rightly argued that these two levels cannot themselves be taken as abstracted from further levels such as biological–evolutionary on one hand, or social–cultural on the other hand. Nevertheless one could probably accept the main message, namely that symbol manipulation and also motivation generation, etc are intimately connected with brain processes (and thus also with each other). This is substantiated among other things through the wide spectrum of relations between malfunctions in cognitive performance and local brain damage – to mention only various types of aphasy, thoroughly discussed also from the epistemological point of view, for instance in 1929 by Ernst Cassirer (see e.g., Cassirer 1990). We shall look for the implications of this reduction hypothesis in the discussion in Sect. 3.3 below.

Neural networks provide a paradigmatic framework for modeling systems with a complicated structure of the state space. Although they are suggestively linked to the connectionist paradigm, they do in fact go beyond it in the sense in which models are typically "less rigid" than theories. Especially in recent decades developments in neurophysiology, on one hand, and in the predictability and characterization of artificial networks, on the other hand (the latter mainly due to the applications of methods from the physics of complex systems), have permitted a systematic framework to be established incorporating both functioning and learning aspects. It turns out that for a very wide range of questions the neural network ansatz provides a basis for identifying structural principles, a fact which we should like to call "the naturalness" of this approach.

3.2   Some Features of Neural Network Models

The Problem of Modeling for Neural Networks A model should not be understood as an approximation to reality. A model is a simplified and artificial "reality", established in the framework of some more or less closed conceptual system and open to our enquiry. It can be an approximation to some other, more refined model but to *that* reality "which one can kick and which kicks back" it can only refer with the help of some conceptual (symbolic) system. If a model is established in the framework of a theory it contains less structure than the theory, but it may be more transparent than the theory itself and also show a certain freedom toward this; moreover, it can also be established in the absence of a theory. In a sense models may be more general than specific theories, since they do not need to explain everything and therefore can constitute bases for building theories.

The problem of modeling amounts to defining those aspects one wants to model and clearly and appropriately limiting the scope and implications of the results.

There are at least three main perspectives from which one can view neural network models:

- that of neurophysiology: as models for brain processes;
- that of artificial intelligence: as models for cognitive processes;
- that of physics: as models for connectivity and for collective phenomena.

Typical modeling problems are to:

- reproduce associative structures between complex stimuli;
- classify, memorize, and retrieve patterns;
- develop optimal behavior under complex constraints.

Most problems, however, have in common, or are even definable in terms of, the question of *storing and retrieving patterns*. Therefore this question is representative for all discussions concerning the capabilities of neural networks.

In establishing a neural network model we must define:

1. the architecture, i.e., the type of units (threshold neurons, linear response neurons) and their connections (directed or symmetrical synapses, etc.),
2. the functioning, i.e., the rules controlling the activity of the network, the integration of the synaptic contribution and the threshold at which, as a result, the post-synaptic neuron becomes active,
3. the learning algorithms (or training), by which the desired response or behavior is ensured by adequate modification of the synapses.

If one sees the rules for functioning (information processing) as a "program", then one can say that artificial neural networks (ANN) are essentially self-programming devices: they develop their own programs according to the requirements to which they are subjected during training. Training just means that the network is asked to do a certain job; then learning algorithms lead to the modification of the synapses as a result of the activity of the network, either depending on the result (supervised learning) or not (unsupervised learning) – according to the specific problem. This applies even to the architecture of the networks: evolutionary types of algorithms have been developed, which allow the training to modify not only the weights but even the size and the hierarchical structure of the network – see, e.g., the chapter of Menzel.

Considerations from Biology and Cognitive Psychology Besides teaching us some general conceptions, like prodigality and like non-linearity and feedback and their evolutionary role, biology provides a number of concepts directly relevant to the neural network approach. Still at the general level, one such concept has to do with understanding the role of fluctuations as a stabilizing effect. Adequacy and survival (or persistence) of complex structures is directly related to the existence of large neighborhoods of viable fluctuations capable of ensuring stable basins of attraction for these structures. This aspect is inherent to neural networks.

Very much information about the brain and its activity is provided by modern neurophysiology – see, e.g. Kandel, Schwartz, Jessel (1995). To a certain extent, this information is incorporated in the neural network models, therefore we shall not enter here into further detail. We want to stress, however, the following observation from neurophysiology: namely, that *for practically every problem there seem to be at least two — but typically many more — mechanisms or implementations*. We thus find:

- mostly threshold, but also linear response neurons (and for the former the threshold behavior is also only approximative);
- local and distant, bundled, divergent or convergent connections;
- non-modifiable and modifiable synapses, with the latter showing non-associative and associative modulation, short term and long term potentiation/desensitization;
- excitatory (mostly axo-dendritic), inhibitory (mostly axo-somatic) and modulatory synapses (axo-axonic: local modulation from a third neuron of the synapse from the second to the first neuron);
- slow and fast synapses (in fact, a wide distribution of synaptic delays);
- electric and chemical synapses, three important types of ion channels among very many others involved in the synaptic transmission, voltage-gated, transmitter-gated, and gap-junction-gated channels;
- biased topological mapping intermixed with functionally determined arrangements of the neurons (e.g., visual field mapping and orientation columns in the visual cortex);
- various hierarchies and group structures with alternate interconnectivity and projection from one group into another, sometimes unidirectional, sometimes with feedback;

and this list is far from exhaustive.

From the point of view of modeling it is very important to try to assess correctly which part of this variability is functional, which part represents stabilizing fluctuations (and therefore in a different sense is again functional) and which is uncontrollable, mostly harmless "noise". Indeed, if the model neglects functional aspects it may miss essential features and if it includes too much contingent variability it gets burdened by irrelevant aspects which may not be harmless in a model which is clearly vastly simpler than a brain. Some clues to these questions can be obtained from the ANN simulations, just by determining the stability of the results under various modifications. On the other hand, the neurophysiology can also offer some indications, e.g. from observations about abnormal behavior.

A cognitive science view is to consider ANNs as modeling "the micro-structure of cognitions" (Rumelhart and McClelland, 1986). This is not to be thought of as providing the ground level of one unique hierarchy of cognitive processes, but as representing background "mechanisms of cognition" in the various cognitive activities. This makes NNs adequate for modeling associative recognition of visual patterns (with the neurons coding the pixels), as well as for modeling semiotic processes, e.g., learning languages (with neurons coding words). The basis for this view is sought by trying to anchor cognitive concepts to fundamental features of the neural networks – it seems, however, difficult to reach stringency here. A better defined question is, how successful are NN models in one or other cognitive problem and what can be learned from the results. In this connection the interesting features of neural networks are:

- the "self-programming" aspect mentioned above,
- the inherent associative character of their information flows,
- the natural way of accounting for multiple simultaneous constraints,
- fuzziness,
- stability with respect to small disturbances,
- easiness of adaptation to new requirements, etc.

Of course, there are also disadvantages ensuing from just these advantages, e.g., the statistic character of their performances; but one of the weakest points seems to be the difficulty in explicitly incorporating high level cognitive procedures.

The Physics of Neural Networks As already noted above, the present development of neural networks is to a large extent due to their becoming accessible to concepts and methods of statistical physics. The reason for this, besides the fact that NNs represent typical systems with many degrees of freedom, with simple units and high interconnectivity, is that the questions one wants to have answered are often in some sense of statistical character. If we think of biological networks, say, each network (brain!) is involved in its own set of similar experiences and confronted with its individual set of similar problems. For questions of cognition we are not interested to find out which *particular* set of patterns can be stored by which *particular* network, but, say, which average number of more or less randomly occurring patterns can be stored as a function of some general characteristics of a class of networks. But these are essentially the kind of questions which statistical physics puts and for which it has developed powerful methods. This has allowed an escape from the original dilemma of the perceptron approach, namely that only simple questions could be answered. Computer simulations, which are well established in statistical physics, are also helpful by providing a kind of "intermediate phenomenology" useful both for the heuristic interpretation and for testing.

Notice that the statistical physics approach in fact generalizes the formal aspects in the functioning of the networks, from the simple propositional logic level already discussed by McCulloch and Pitts to the complex mathematics of state spaces and fixed point flows. Paralleling McCulloch and Pitts' stand, the artificial networks are understood as modeling those formal aspects of the nervous activity which are independent of the particular physical and chemical structure: *We search for general structural and functional features, the development of which is dependent on the history, activity, and experience.*

Most analyses in the statistical mechanics of neural networks concern the "Hopfield model", which describes an interconnected net of 2-state neurons. The basic functioning is the summation of the impulses coming via synapses to each neuron so as to trigger its discharge if a certain threshold is achieved. The principal type of problems relate to the structure of the state space of this model and can be phrased as a question concerning the capacity for storing and retrieving patterns. There are two main parameters of the model, at least in the simplest (homogeneous) architecture: one structural (the "connectivity" n, the number of synapses made on the average by each neuron) and one functional (the "temperature" T, the amount of fluctuation in the reaction to the presynaptic potential). The state space structure can be understood with the help of an energy function and depends discontinuously on the parameters, reflecting phase transitions. A specific result is that for storage by optimal training algorithms the capacity of retrieval is controlled by the ratio $\alpha = p/n$ with p being the number of patterns to be stored. Explicitly, there exists an $\alpha_c(T) < \alpha_c(0) \simeq 0.14$ above which storage and retrieval capacity disappears, while for $\alpha < \alpha_c$ this capacity exists and shows special features depending on the temperature.

A second class of problems concern the learning algorithms. The biological basis for all learning algorithms is the synapse plasticity revealed by the so called "long term potentiation" (LTP). Its associative realization is abstracted in the Hebb rule: synapses which contribute positively to a firing neuron (which are "on the winning side" so to speak) become strengthened. The results of the analysis of various learning algorithms concern first general convergence proofs. Secondly, general proofs have also been obtained for combined learning and activity dynamics with well separated time scales (slow dynamics for the learning and fast dynamics for the functioning – a biologically plausible situation). Thirdly, more complex situations have also been studied, e.g., the role of "unlearning" in discarding spurious patterns, or the learning of conflicting associations (realizing alternatives). In these cases general proofs are usually missing but there are a lot of data available from simulations.

Finally new developments put the question of non-trivial time structures. On the one hand one wants to implement the synapse delay structure observed in biological systems and obtain in this way memorization and retrieval of time series. These are realized as limiting cycles of the activity and typically involve asymmetric synapses (again, biologically justified). On the other hand a more fundamental question is whether only the average firing rates carry information or whether there exist well timed events in the brain, that is, whether the timing of spikes is itself relevant. While there is little secure biological information about the latter question, strong motivation for studying such aspects comes from the cognitive problems known as "binding": How do we recognize separate features in a complex structure ("pattern segmentation") and how do we realize that a number of separately recognized parts build one complex pattern ("feature linking")? Notice that the study of non-trivial time evolution in fact needs an enlargement of the framework provided by statistical physics – which essentially deals with equilibrium phenomena.

3.3   Questions Concerning the NN Approach

The neural network ansatz touches upon our understanding of reductionism in biology and cognition, the mind/body problem, the concept of organism, and other questions too.

Let us first briefly illustrate its relevance, say, for our concept of an organism. We have already noticed the role of fluctuations in ensuring basins of attraction and thereby stable evolution of complex structures. This question is not essentially different in species evolution or in individual learning. From the experience with neural networks we can obtain hints for a quantitative understanding of the basin structure[2]. This suggests that accounting for fluctuations in establishing the concept of organism leads to well-defined, quantitative points of view.

Let us now consider the question of cognition, and let us recall the reductionist statement of McCulloch and Pitts quoted in Sect. 3.1. It appears that what we observe in the NN approach is a relativisation of the dividing line between what we may call "brain processes" and what we may call "mental processes". In fact we seem to be able to model *certain* basic features of very different kinds of cognitive behavior at all levels in the framework of the neural network approach.

We have used in this connection the word "naturalness" in a practical sense. However the word suggests a deeper naturalistic point of view which surely is characteristic to the whole of the "neuroscience" and "neurophilosophy" approach. We shall now try to make some comments concerning this problem.

A first question is whether this approach has brought forth a "shift in vocabulary" which has trivialized the claim that "mental states are brain states", such that a question like "are mental states part of the physical world?" could only be answered "what else?" – Tetens (1994). Let us first try

---

[2] For instance, one can relate (Gardner 1988) the estimation of the basin of attraction defined as the ensemble of patterns which all have the same attractor, to the evaluation of the sensitivity of this attractor to changes in the synaptic structure.

to establish the sense of the latter statement. When we consider the "mental states" of somebody, they are for us part of the "physical world" since we have to take them into consideration at the level of physical interactions. Then, since "[...] to be the Other is neither an accident nor an Adventure which may or may not occur to a human being, but it is from the beginning one of his attributes"[3], we must admit that the same holds true for our own mental states. But this immediately implies that we must include society in this world and consider culture just as one of its "dimensions". This just means that such pairs as "spiritual/physical", "mind/body", "subject/object" should not be hypostatized but applied as instruments of reflection. To belong to the "physical world" is part of the truth concerning "mental states".

In a very simple sense mental states are states of the brain, since nobody can think without a brain, but one can think at any given moment without anything else. But this does not mean that this is enough for defining, understanding, or describing mental states. For this we surely need to take into account the communicative, cultural level. Therefore, following e.g. Putnam (1992), mental states are not reducible to neurophysiology. Can we consider this merely as the failure of a "methodological reduction", a question of practicability? If we develop the argument of Putnam then this is not the case. The problem resides in the impossibility of *achieving* mental states without inter-subjective communication: Even if at any given moment mental states *are* realized as brain states they cannot be separated from their "history" and therefore they always include a reference to previous inter-subjective interactions. This also means that connectionism and symbolic manipulation must collaborate – possibly together with other aspects – in a description of mental phenomena.

We are thus led to ask about the character of this "collaboration". Such a "collaboration" does not appear arbitrary and hence could hardly concern totally independent structures. The observations from the neural network approach suggest that the following interpretation of the McCulloch and Pitts' conjecture might offer an answer to this question. The immanent aspects of the brain activity, which themselves have been developed during a long evolution in constant interaction with "the world", will transpire in all the structures based on the brain activity at all levels, including the historically evolved cultural structure. Independently of how these aspects are realized at each level, they will always provide a kind of compatibility and interdependence between the structures at these different levels which will ensure the collaboration between them. On the other hand both evolution and learning will allow a feedback of the "higher level" structures down to the elementary level of the synaptic – hence, neurophysiological – structure.

---

[3] Ortega y Gasset (1995); ad hoc translation.

References

Cassirer, E. (1990): *Philosophie der symbolischen Formen, III* (Wissenschaftliche Buchgesellschaft, Darmstadt)

Churchland, P.M. (1989) *A Neurocomputational Perspective* (MIT Press, Cambridge MA)

Gardner, E. (1988): "The phase space of interactions in neural networks", J. Phys. A: Math. Gen. 21, 257

Kandel, E.K., Schwartz, J.H., Jessel, T.M. (1995): *Essentials of Neuroscience and Behavior* (Appleton and Lange, Norwalk)

McCulloch, W.S. and Pitts, W.A. (1943): "A Logical Calculus of the Ideas Immanent in the Nervous Activity", *Bull. Math. Biophys.* 5, 115.

Ortega y Gasset, J. (1995): *El hombre y la gente* (Alianza Editorial, S.A., Spain)

Putnam, H. (1992): *Representation and Reality* (MIT Press, Cambridge, MA)

Rumelhart, D.E. and McClelland, J.L. eds, (1986): *Parallel Distributed Processing: Explorations in the Microstructure of Cognition* (MIT Press, Cambridge, MA)

Tetens, H. (1994): *Geist, Gehirn, Maschine* (Reclam, Stuttgart)

Intelligence: A Nomadic Concept*

Clairette Karakash

Institute of Hermeneutics and Systematics, University of Neuchâtel,
CH-2006 Neuchâtel, Switzerland. e-mail: clairette.karakash@theol.unine.ch

The first attempts to automate human intelligence date from the fifties. Three factors had to be combined to make the undertaking at all possible: A concept, a tool, and a language. The concept or logical principle was first stated by A. Turing (1937), and may be summarized as follows: Any formalizable process can be reproduced by a machine capable of performing an ordered series of operations on a finite number of *symbols*. The tool was supplied by J. von Neumann, the inventor of pre-recorded programs[1], the predecessors to modern software systems. All that remained to be invented for digital computers to be capable of processing non-numerical symbols was one or several languages intermediate between a binary system and natural language. Once these conditions were satisfied, the hope was to automate any operation of the human mind, assuming that the latter could be described as a series of operations leading to the performance of a task. Attempts were therefore made to computerize problem-solving, theorem proofs, games of strategy, interlinguistic translation, and shape recognition.

Forty years after the first symposium on artificial intelligence[2], what conclusions can we draw from the experience gained in the field? Although no machine is yet capable of passing the Turing test, we can nonetheless measure the progress made: We now have powerful tools for design purposes, learning tasks, and manufacturing processes, as well as an impressive array of highly effective software systems in a certain number of specialized areas. We have access to a vast quantity of collective memory banks and knowledge trees, all available at a moment's notice and regularly updated. Computer-assisted communication enables not only real-time processing and transmission of information, but also participation in the virtual world of groupware systems and planetary networks such as Internet, Swift, Sita, Renater etc. The feats of which computers are capable today far exceed anything the scientists that met in Dartmouth in 1956 might have imagined even in their wildest dreams.

---

* The term *nomadic concept* is taken from the title of a work published by I. Stengers (1987).

[1] Von Neumann first had the idea in 1943, when he was working as a consultant on the construction of ENIAC, the largest computer of the day. He presented his automaton theory at the Hixon symposium in 1948, but only published it in 1966.

[2] At Dartmouth College in the summer of 1956, J. McCarthy gathered together the world's finest logicians, computer scientists, and cybernetics technicians involved in the design of automata, to launch the idea of artificial intelligence.

Machines are increasingly capable of simulating human reasoning and behavioral patterns; unlike the forefathers of artificial intelligence (AI), however, we are far from considering computers as being potentially intelligent. The paradox is worth examining in closer detail.

Curiously enough, the notion of natural (human) intelligence has turned out to be dependent on AI. The proof is that progress in the latter constantly incites us to redefine the former. In the AI mirror, the concept of intelligence has become labile, as if the object of our study changed with our research. The purpose of this article is to seek out the reasons for this state of events, and to meditate on the transformations that AI stamps onto the understanding human beings have of themselves[3].

1   Lability and the Concept of Intelligence

The concept of intelligence changes and becomes more complex during its study

As a result of tests in intelligence automation and, of course, due to the development of cognitive science, we know a great deal more about the ways in which the human mind operates than we did 40 years ago. Initially, intelligence was equated to a rigorous form of reasoning. By trying to computerize games of strategy, theorem proofs or problem-solving, we came to realize that formal logic was not in itself sufficient to account for human performance. To play or think intelligently, human beings combine positive thinking with empirical reasoning. We base our acts on experience and on knowledge; we evaluate, we deliberate. In order to improve their software systems, design engineers took for their model human specialists. Expert systems include a knowledge base, their inference engine combining algorithmic rules with heuristic procedures. Today, these computers can match the world's best chess players or perform diagnostic analyses in highly specialized fields.

Attempts to automate translation have shown that it is not enough to give a computer a lexicon and a grammar for it to produce correct sentences. Contrary to the beliefs of the structuralists, language is not a closed system; to understand it one needs more than simply a code. To translate, one must understand, and to understand the meaning of sentences a relation must be established between speech and the extra-linguistic world, between words and things. How is meaning corroborated by signs and groups of signs? This is the famous question posed by Searle, who invented the Chinese-room argu-

---

[3] We have no intention here of venturing out onto quicksand with conjectures about the limits of AI: *A priori* judgements have too often been proven wrong in the past for us to ignore our forebears' experience! Out of intellectual rigor and consideration for the efforts of scientists, we shall limit ourselves to assessing what AI has changed in our understanding of the human being and in the collective mind-set.

ment (1980) to highlight the fact that computers process information at the formal level, without understanding the meaning of the symbols they compute. Searle was no doubt right in saying that the semantics are injected from the outside by the programmer. Computer scientists, however, did not ask the machine to generate meaning. They thought that by introducing logical, correctly formulated expressions into the machine, it would be possible to automatically transform them into other equally logical and correctly formulated expressions, using a certain number of lexical and grammatical rules. In this case, the obstacle to the undertaking was not combinatorial exponentiation, but the polysemy of language. We all know words and sentences can have several meanings. How do we choose between the various interpretations possible? Based on research work in pragmatic linguistics in particular, we now know that the meaning of a sentence depends on the context of utterance. Yet in most cases the context remains implicit, since the speaker supposes his listeners or readers are as aware of it as he. If computers produced nonsense, it is because they had no implicit knowledge of the world, what we call common sense, which human beings obtain through education and experience. During the seventies, efforts were therefore made to supply the machine with structured knowledge of a general kind, in the form of semantic networks, conceptual dependency diagrams, scripts and scenarios, plans, and so on[4]. Common sense, however, is much more difficult to formalize than rigorous reasoning. Projects which require too extensive a knowledge of the world have had to be abandoned, and subsequent efforts restricted to modeling knowledge and know-how in clearly defined areas. Automatic translation is used for weather forecasts, for instruction leaflets issued with drugs or appliances, in short, in areas in which language is relatively technical.

During the tests mentioned earlier, our hypotheses about the operation of the human mind had to be reviewed. Intelligence cannot be reduced to a series of logical operations; it implies knowledge and skills. Intelligent behavior presupposes the ability to perceive the context, to learn, to benefit from past experience, and to assess the immediate situation. The concept of intelligence has thus become considerably more complex, to the extent where it now encompasses not only cognitive faculties but also sensorial and motor faculties (Agre, 1988, van Gelder and Port, 1994). Intelligence is now seen to be "the art of acting in, on and with the world" (Ducret, 1991). According to Minsky (1986), intelligence should be plural, and the human mind analogous to a multi-agent system. The on-going re-definition of the phenomenon under study makes modeling it all the more difficult. This is why AI has developed in a number of directions, which can be categorized under two main headings, depending on the purpose:

---

[4] We refer here to the semantic networks of Quillian (1967), Minsky frameworks (1975), and the scripts or plans of Schank (1977). For the difficulties encountered in automated translation, see Steiner, Schmidt and Zelinsky-Wibbelt (1988). For progress in the processing of natural languages, see Waltz (1989).

- modeling performance, i.e., arriving at the same result as human beings, no matter what procedures are used;
- modeling human approaches, i.e., formally reproducing the underlying mechanisms.

The two areas in fact cannot really be separated, since AI has confused all the traditional interdisciplinary boundaries; not only between theoretical and applied research, but also between the natural sciences, the humanities and philosophy. AI has become transdisciplinary, nomadic if you will, its followers stretching out behind as it migrates from one territory to another, with conceptual or technological breakthroughs in one area providing feedback for all the others. The functional interaction described here is twinned with recursiveness on the epistemological level.

2 The Epistemological Loop: A Vicious Circle?

To model intelligence requires an understanding of human beings;
to understand human beings requires intelligence

2.1 The Initial Postulate

The founding hypothesis of AI is that the human mind enjoys a certain degree of independence in relation to the material of its natural support, the brain. The postulate of *material independence* is shared both by supporters of the *computo-symbolic* approach and by those who favor the *neuro-cybernetic* approach[5].

In the computo-symbolic approach, the cognitive system is described as an information processing system. Reasoning is equivalent to computing, i.e., to manipulating a set of symbols on the basis of pre-defined rules. The computational theory of the mind postulates the existence of mental representations encoded in the neuronal architecture in the form of signs, or 'tokens'. By definition, a system of tokens has physical and semantic properties. It can therefore serve as an interface between the neurophysiological and cognitive levels. The brain is to the mind what hardware is to software. The postulate of material independence is twinned here with the principle of formal equivalence (Haugeland, 1985). This approach has enabled the automation of most of the logical procedures carried out by the mind, from games of strategy to computer-assisted design, via computerized expert systems, weather forecasting, and currency exchange rate projection.

The *neuro-cybernetic or connectionist* approach also supports the postulate of material independence, but proposes a functional architecture different to that of von Neumann, namely interconnected networks of neuronal

---

[5] For the origins of AI and its relation to cybernetics, see the work by S. Heims (1991): *The Cybernetics Group*.

automata, which use digital information and work in parallel. Impulses from the surrounding environment are transmitted by propagation throughout the system. There is no central control unit: Instead, all the units gradually adjust in relation to each other based on an activation rule, until the system reaches a steady-state or a limit cycle. The global state obtained constitutes the reply by the system to the stimulus from the environment. From Rosenblatt's Perceptron to the networks of Hopfield and Kohonen, from Boltzmann's parallel machines to networks based on parallel distributed processing, the variety of networks is considerable, and their architectures differ in relation to the tasks involved (Amit, 1989; Bourret et al., 1991). All of these are dynamic, adaptive models. The connectionist approach has obtained some satisfactory results precisely where the computo-symbolic approach had failed: In shape and speech recognition, in associative memory, and in supervised, non-directive, and reinforced learning tasks. In addition, networks are strong tools when confronted with local perturbations or deterioration.

Although they are antinomic, the two approaches are probably complementary, but there is no widespread agreement about the way in which they are articulated. For some, the computo-symbolic approach attempts to apprehend the most abstract level of intelligence, while the connectionist approach concerns itself with describing the microscopic, sub-symbolic level (McClelland and Rumelhart, 1986; Smolenski, 1988; Clark, 1989). For others, connectionism includes conventional AI as the borderline case (Varela, 1986)[6]. The connectionist approach was for many years eclipsed by the computo-symbolic paradigm. Since the beginning of the eighties, however, the latter has seen something of a revival, due to the progress of computers on the one hand, and to the rise in fluctuation-based order theory and progress in the neurosciences on the other (Grimson and Patil, 1987; Andler, 1992).

2.2   Methodological and Epistemological Recursiveness

In order to implement intelligent procedures by computer, our starting point has to be based on hypotheses about the operation of the human mind. It must be possible to describe a reasoning process, a strategy or any other operation of the intellect in terms of reasoning steps in order to be executed by a computer. The aim is to break down and formalize the processes that we wish to automate. When the results obtained do not coincide with the expected results, the scientific approach consists in rectifying the stated hypotheses about the dynamics of the phenomena under study. The approach presupposes that the phenomenon in question has not been modified by the experiment. Yet human intelligence seems to be transformed as a result of contact with AI, unless it is only our concept of intelligence which changes. How can we distinguish between the two alternatives, however, when human

---

[6] For an overview of the debate between the cognitive and connectionist schools, see Reeke and Edelman (1988) or Memmi (1990).

behavior patterns change as a result of contact with automata, as can be seen today on a large scale (see Sect. 3 below)? If intelligence is an anthropological constant, the trial and error approach is wholly justified: With patience and perspicacity, we shall succeed in describing the phenomenon with increasing precision, much as we have done for the stars or the genome. On the other hand, if our investigation affects the object under examination, we find ourselves in a situation comparable to that of quantum mechanics, in which the experiment modifies the characteristics of the object under study. Even in the latter case, however, the situation is not totally hopeless, since the theory can take the interference into account. But the analogy between the current situation in AI and that of quantum physics is only partial. In the latter the experimental set-up is not modified by its interaction with the object under study, whereas when modeling intelligence, *the object under study is at the same time the tool of investigation*: We have to use intelligence to clarify the mechanisms of intelligence! Which means that if the object changes as the study progresses, the instrument does likewise. We are faced with a situation of recursiveness, both methodological and epistemological. Amongst other things, this dual circularity raises the question of how to validate the procedures used and what status may be given to the knowledge thus obtained.

2.3 Validating AI Procedures

How can we evaluate the pertinence of theoretical hypotheses and the performance of software systems, if the former serve as models for the latter and vice-versa? In this recursive situation we have a single equation in which both values are unknown. Without the independent data supplied by the cognitive science and the neuroscience there would be no solution, and the circularity mentioned earlier would simply be a vicious circle. Such data are now widely used by AI research workers. The following are two examples:

- In the processing of natural languages, dynamic programming between lexical, syntactic, semantic and pragmatic levels has been introduced as a result of the observations made in psycholinguistics and neurolinguistics.
- The modeling of elementary cognitive processes is covered today by robotics, since genetic psychology has shown that perception and motricity are decisive factors in the development of intelligence[7].

---

[7] Mention should be made in this respect of recent work by the Computer Science and Artificial Intelligence Institute of the University of Neuchâtel, which continues the phenomenological tradition and fits into a constructivist epistemological perspective. See Müller and Rodriguez (1995), Faihe et al. (1996), Müller and Pecchiari (1996). Mention should also be made of the work at the microcomputing laboratory of the Federal Polytechnic School of Lausanne: Gaussier and Zrehen (1994), Zrehen (1995). In both these research centers, robots navigate without a map of their environment, and orient themselves with sensors.

Because of this input of exogenous data, the system is open, by integrating information from the outside. Modeling takes place in spiral and not in closed circular form. The process of gradually rectifying the data and their meaning is similar to the hermeneutic circle of theological and philosophical tradition. This at least is what was put forward by the artificial intelligence laboratory at the MIT (Mallery et al., 1986) in comparing the recursiveness of textual interpretation to the bootstrapping used in AI to update knowledge bases (Davis and Lenat, 1982; Haase, 1984). The term *bootstrapping* is understood to mean "a process which uses a lower order component (a *bootstrap component*) to build a higher order component that is used in turn to reconstruct and to replace the lower order component". This in fact is a procedure often used by the human mind to deduce the units of meaning of a text or to fine-tune the information it conveys. The comparison would have been surprising ten years ago, but the current revival of hermeneutics (which was as ostracized as neurocybernetics) suggests that the analogy should be examined in closer detail.

From F. Schleiermacher to U. Eco, the theoreticians of textual interpretation have highlighted the fact that understanding is a gradual process: Cursory reading gives a general first impression, which is slowly fine-tuned with further readings. Approximating the meaning of the text is obtained in concentric circles, from partial understanding to global understanding and vice versa, while observing the principles of identity, non-contradiction and overall coherence. The interpretation depends not only on the sentences and their potential meanings, but also on the reader and his presuppositions and knowledge, both of which play a decisive role in comprehension: The reader's world view, his relationship with the real and virtual worlds all modulate his overall understanding of the text. In addition, when the context in which the text was produced does not correspond to that of its interpretation by the reader, relations of equivalence have to be established from one context, or world, to the other. In other words, the context and its interpreter are one and the same system. The meaning emerges from a spiral progression between these three instances or parameters. Since the interpreter is a participant fully involved in the process of constructing the meaning, there are many interpretations possible. This indicates the extent to which interpretation is open-ended. This concept of gradually constructed units of meaning contrasts with that of the unidirectional transmission of information encoded by an emitter and decoded by a passive receiver.

Another example of recursive interpretation is that of medical diagnosis which operates on non-linguistic signs, namely physical symptoms. In this example, two subjects, the patient and the doctor, have as their data a certain number of symptoms that have to be interpreted. Initially (first circle), the doctor listens to the patient, before attempting to place the information received into a context by asking questions and establishing the anamnesis. This leads the doctor to put forward an initial series of hypotheses, which he then tests by examining the patient (second circle). He will confront his hypotheses with the results of the examination: He evaluates the overall result, which probably leads him to fine-tune his hypothesis. Where necessary, he will attempt to obtain further information, by proceeding with clinical tests and taking into account the laboratory results (third circle). On the basis of this overall information, he will finally make his decision. In other words, he has interpreted the symptoms using the method of differential diagnosis. With each spiral of this "hermeneutic circle", the doctor will have used his intuition and his knowledge base, both of which are modified and enhanced with experience.

Generally speaking, systems of knowledge are constructed recursively. To take another example: The successive theories on the atom (Gassendi, Lavoisier, Rutherford, Bohr, etc.) suggested a certain number of experiments, which in turn modified our representation of the elementary structures of matter, and produced new theoretical models. The same reasoning can be applied to our understanding of cultural or social phenomena. AI therefore uses a process generally employed by the human mind, a process of self-organization, on the basis of which the cognitive system becomes more complex.

It is the integration of external data which guarantees the openness of the system and enables it to evolve. From the epistemological point of view, the question of validating its results has simply been relocated, however, since what ultimately guarantees the validity of the external data? This means tackling an issue which has been hotly debated for many years in philosophy, that of the status of knowledge and its legitimation. The issue is all the more crucial in that many research workers in the cognitive science and neuroscience have considered the brain as a data processing system and adhered to the computo-symbolic paradigm.

2.4   The Status of Knowledge in AI

Over and above the question of procedures and their validation lies the issue of determining on what the legitimacy of our knowledge systems is ultimately based. Using as their starting point the postulate that the world exists independently of our way of perceiving it, the realist and positivist epistemologists presupposed the existence of a reality *per se*, distinct from its perceiver. Putting their faith in the power of human reasoning, these epistemologists, K. Popper included, postulated that with sufficient perspicacity, science was capable of providing an image of reality which was increasingly compliant with that reality, sufficiently compliant in any case for us to consider scientific theories and the knowledge systems they entail as a reliable, albeit approximate, form of knowledge of the world.

In the first half of the twentieth century, a series of discoveries threw these certainties into turmoil: The advent of quantum mechanics, of Gödel's theorem, Wiener's cybernetics, Shannon's information theory, and the work by von Forster, Piaget, and Bateson, to mention but a few of the factors which caused our foundations to crack. The fact had to be faced that science is a construct of the mind just as history, moral values and economics. No doubt for these different fields the object and the rules of the game are different. But knowledge always depends on the way in which the mind breaks down and structures reality, as was argued in exemplary fashion by Atlan (1986). Even if one does not adhere to the sociological arguments of T. S. Kuhn, to the anarchism of P. K. Feyerabend or to the radical constructivism of P. Watzlawick, one has to recognize that where validating scientific knowledge is concerned, the postulate of objectivity has generally given way to intersubjectivity.

In the light of research on the structures and modalities of knowledge, in fact, all legitimising principles have turned out to be contingent. This is why constructivist epistemology has replaced realist and positivist knowledge theories. The common denominator to constructivist points of view is the integration of the notion of a *subject*: "Knowledge implies a cognizant subject and has no sense or value without him" (Le Moigne, 1995, p. 67). Knowledge is no longer considered to be an objective truth, independent of the agents and modalities of cognition. Knowledge is seen to be the reflection of the cognitive experience, i.e., the interaction between a cognizant subject and his surrounding environment. From that point onwards, the legitimacy of a knowledge system is no longer measured in terms of its degree of congruence with reality but on its pragmatic value, in its ability to enable an action or behavior suited to its environmental constraints. From this epistemological point of view, knowledge systems are mutually supporting, a fact which is particularly obvious in the field of AI. Theories are bench-marked in relation to experiment and only become credible if their performance is recognized by the scientific community. Ultimately, deciding that a machine or human behavior pattern is or is not intelligent, is a question of intersubjectivity, i.e., a mixture of beliefs and knowledge systems about intelligence collectively shared in a given culture and time. In no way does this affect the efficiency nor the rationality of the knowledge systems themselves; it merely changes their status in our minds. This makes it easier to understand, however, why in epistemological terms intelligence is a constructed concept, which does not mean it is arbitrary, but nomadic, and therefore difficult to grasp.

To say that knowledge is a construct of the mind implies that the practitioner must be taken into account in knowledge theories. The subject in question, however, is not the transcendental Kantian kind, as J.-J. Ducret (1992, p. 23) so aptly pointed out, while underlining the fact that constructivism revises the thesis of the *a priori* nature of forms of understanding[8]. At its birth, a baby apparently makes no distinction between self and non-self. It

---

[8] On the other hand, neither classic AI nor cybernetics had the initial purpose of producing a cognizant practitioner as their initial purpose; the aim was.rather to determine the formal constraints which have to be met in order for a machine to accomplish intelligent tasks. See Dupuy 1994, pp. 112–116 and Proust, 1987.

gradually discovers itself as distinct from the environment. Subject therefore does not pre-exist object, as some form of pure transcendental conscience. The autonomy of the subject is a property acquired through experience, through physical interaction with the environment and linguistic interchange with other human beings. Knowledge is not a mirror of the world (Rorty, 1981); it results from the experience an individual has of reality, and then influences the way we construct and organize that knowledge: The cognitive structures are developed step by step via interaction with the environment. Genetic epistemology has shown that becoming an autonomous subject and constructing reality are interdependent.

— *"Intelligence (and therefore the act of knowing) thus does not start with knowledge of oneself, nor by knowledge about things as such, but by knowledge about their interaction; by simultaneously orienting itself towards both poles of that interaction, knowledge organizes the world by organizing itself."* (Piaget 1937, p. 311)

If intelligence is both the cause and the effect of the knowledge the individual has of the world and of himself, this forces us to reconsider the distinctions traditionally made in our modern world between subject and object, between being and doing, between autopoiesis and heterogenesis. Since AI has greatly contributed to rendering these three frontiers fuzzy, the last section of this contribution focuses on the consequences that this epistemological change has had on our self-understanding.

3   The Influence of AI on Self-understanding in Human Beings

— *"By changing what we know about the world, we change the world we know. By changing the world in which we live, we change ourselves."* (Dobzhansky, 1962, p. 391)

There can be no doubt that the invention of the computer has considerably changed social practices and the organization of work, giving rise to the sharpest criticism as well as the wildest dreams. One need only compare H. Simon's optimism with M. Heidegger's statements about cybernetics to realize the extent of the passion that has been aroused by attempts to naturalize intelligence. For a number of years, philosophers, epistemologists, and anthropologists alike have tried to measure the changes brought about by the computer revolution. There is no more hesitation about comparing it to the invention of writing or the printing press. We are apparently the contemporaries of a new revolution of Copernican scale and scope, in which the cognizant subject is relegated to the wings in relation to an ecology of the mind, to use G. Bateson's expression, although he was not thinking of a planetary network like Internet, but of the collective intelligence of complex systems such as the family, society, or institutions.

3.1   The New Alliance Between Subject and Object

Kantian epistemology carefully made the object dependent on the subject, so that knowledge of the world may be derived from the categories of understanding, and not from the signals perceived by the sense organs, as did empiricism. With his analysis of perception, Husserl has taken the subjective pole of cognition even further, by emphasizing the intentional character of a transcendental conscience which aims at and captures objects (self-donation of the world by the subject). By studying the development of the structures of cognition, (genetic epistemology) Piaget has rehabilitated the role of the object in the constitution of the subject: The conscience of self is not inherent to our being, but it emerges from interaction with the environment and with other participants. Individualization and autonomy are products of ontogeny and education. This does not invalidate subject-based philosophies since, if knowledge is acquired, the conditions enabling cognition (the structures of understanding) are the result of phylogeny and are, for that very reason, inherent to our species. A great deal of thought has been given to the innate part of human intelligence, but intelligence does not seem to be as hereditary as we had imagined.

In its own way, computer science has taken over from genetic psychology in reconciling object and subject. Like a teddy bear, the computer is another us, a double to whom we talk, with whom we work and play. It is a *virtual subject*, or, to use the terms employed by psychologists, a *transitional object* which enhances communication. In many fields, the computer has become the necessary intermediary between human beings, whether in banks, companies, air navigation, or medical imaging systems etc. As computers and users have become networked, so cognitive systems have emerged both inert and alive, composed of mind and matter, signs and senses. The cognizant subject no longer sees himself as a transcendental being: He merely represents another node in the network which encompasses him, a part of the system just as the tools he created in order to dialogue with the world. In other words, computers are fully-fledged participants in the development and organization of knowledge.

Strictly speaking, the alliance between subject and object is not new: From telescopes to scanners, from sundials to atomic clocks, from tam-tams to fax machines, artifacts have become partners in human cognition. The distinction between thinking subjects and thought objects disappears when learning takes place using interactive teachware, and the transmission of knowledge or know-how requires technical media as a support. Computers may never become as intelligent as human beings, since the latter always becomes more ingenious as a result of contact with tools. What is certain, on the other hand, is that no-one has ever become intelligent on their own, with the use of any tool or educational support. AI has reminded us, indeed taught us, that human intelligence is partially artificial.

3.2   The Synergy of Understanding and Acting

If we take the role of tools in cognition seriously, it entails giving relative value to any philosophical tradition which tends to contrast self-understanding with instrumental reasoning. Over-simplified opposition of this type does not do justice to the development of human comprehension. It is by handling objects that children discover the texture and structure of the things that surround them, and form their personality. It is by progressing from observation to experimentation that modern science has provided itself with a system of efficient knowledge. Thanks to AI, it is now possible to simulate experiments on the Turing machines we have in PCs. The first computer simulation programs date from the fifties, with the airspace monitoring system SAGE (Semi-Automatic Ground Environment). Another twenty years had to pass before simulators integrated the first image generators. Today, we can simulate and display a vast range of phenomena, from material resistance tests to battle plans and surgical operations.

Simulation consists in modeling the dynamics of a phenomenon or an object using a simplified system. The latter is a tool for managing complexity. To model a phenomenon, the parameters that govern its evolution have to be selected. Extracting the relevant parameters in order to develop a model is a complex operation, since it depends both on the intrinsic characteristics of the phenomenon and the purpose of the simulation (computer-assisted design, or decision-aid, computer-assisted learning, research work, quality testing, entertainment, etc.). Computer models enable parameters to be varied at will, and exploration of the arborescence of possibilities both quickly and cheaply, with an efficiency which was unknown before the invention of the computer. Its virtual nature makes simulation similar to thought. Its dynamic nature makes it similar to experimentation. Simulation consists in carrying out a series of actions in a virtual world in order to study their effects. The purpose of modeling is to pilot or facilitate the action, to give it tangible substance in the real world.

— *"Although basically problematic, the virtual world is like a subjective situation, a configuration of trends, forces, aims and constraints which are solved by an actual event, in the real sense of the word. An act is performed which in no way was pre-defined, which in return modifies the dynamic configuration that gives it meaning. The relationship between the virtual and the real is the basis for the very dialectic of the event, of the process, of the* **self as a creation.** *"* (Lévy, 1995, p. 135).

Artificial intelligence has reminded us, indeed taught us, that human understanding is based on action both in tangible and virtual reality. Knowledge evolves as a result of the interaction between the real and the imaginary worlds. Understanding, including ourselves, requires doing: Not only the know-how of doing, but also the know-why, where and when of doing, as well as a great deal more.

3.3   Matching the Intelligible with the Sensible

AI has resurrected the celebrated mind-body issue, which has been debated in schools of philosophy ever since the Ancient Greeks. The powers of imagination and abstraction provides the human mind with a certain degree of autonomy in relation to its substantial, sensible, tangible body. That autonomy is limited, however. All it requires is for a part of the nervous system or for certain vital organs to be damaged, and the mental functions are perturbed. It is trivial to say that sensibility is the key parameter to cognition, since the sensorial receptors form the interface between the individual and his environment, between self and non-self. Yet we are far from elucidating the mechanisms whereby sensations are transformed into perceptions, representations, or ideas. Nor do we know how intent causes action. All sorts of theories have been set up to link neuronal activity to thought (Edelmann, 1992; Penrose, 1994; Varela, 1996). In this contribution we do not address the nature of mental representation, since the first chapter addesses the issue (see U. Ratsch and I.-O. Stamatescu). We shall simply point out a possible effect on self-understanding in human beings of representations artificially generated.

Computer-generated images are a cross between the intelligible and the sensible, since they make an abstraction perceptible. Unlike photographs and video recordings, virtual images are not reproductions of the real world, but constructs of the mind, into which the computer breathes "life", gives plasticity. Ph. Quéau, a world expert on computer graphics, considers computer-generated images as a *new form of writing* which can be animated, transformed at will, and can create meaning. The power of virtual images stems from the fact that they imitate reality extremely well. As the technology becomes perfected, it is increasingly difficult to distinguish a digitized image from a computer-generated image. The difference between reproduction and imitation is slowly disappearing, and we therefore have to be even more careful with what seems to be self-evident.

> — *Images, which have become an omnipresent means of writing, should no longer be taken for granted, or absent-mindedly* **seen**, *but must now be carefully* **read**, *analysed, and compared with their context, as we have learnt to do elsewhere in the field of written information.* (Quéau 1993, p. 37).

A spectator sitting in his armchair watching three-dimensional images displayed on a stereoscopic viewer has the physical and emotional reactions initiated by the illusion of moving and acting in the virtual world. The fires and thunderstorms simulated on-screen may not burn or wet him, yet the activation of his sensorial receptors is capable of misleading his reason, even though the spectator is perfectly aware that he is watching a work of fiction: Such is the mysterious matching of the intelligible and the sensible. Like astronauts who stumble when they return to earth, cybernauts, the travelers of virtual space, do not return to the real world unaffected. Their immersion in the world of make-believe offers them a glimpse of possible other worlds, sometimes totally different from the real universe. The laws which govern daily life are no longer applicable, the mind can surf in absolute freedom, simultaneously adopting several points of view, playing various character roles, navigating through time, building and destroying empires, in short, recreating the world in a number of ways. Fiction favors what P. Ricoeur (1986, p. 129) calls "imaginative variations of the ego": Settings for the self that vary in relation to the scenarios offered. This virtual reality of the person thus provides various viewpoints for understanding oneself, suggesting implementations of the possible world in reality. Its creative power is based on the alienation of reality and the removal of the precepts of common sense.

The perspectives opened up by computer graphics are not wholly new. Fictional novels, fantastic literature and science-fiction, films and video games all have similar effects. The interaction, however, between the senses and the imagination are multiplied by computer modeling, enabling the spectator to become an actor in virtual reality simply by using a keyboard or joy-stick. This may not solve the mind-body issue, but AI means we cannot neglect the senses in the creation and operation of human intelligence. Reciprocally, it draws our attention to the role of imagination and abstraction in self-understanding, even if we must sometimes beware of hallucinations.

3.4   The Connection Between Local and Global Intelligence

In Renaissance Europe, scholars traveled and exchanged their writings far and wide, thanks to the recent invention of the printing press. Much later, the telegraph and then the telephone enabled the remote exchange of messages without requiring written media. Yesterday with the fax machine and today with E-mail we can instantly transmit texts and images over the whole surface of the planet. Television constantly pours images and voices from around the world into our living rooms. AI goes even further: It provides us with the right of reply in real time. The networks that interconnect individuals via computerized machines enable them to communicate, work, trade and play together. The virtual communities they thus create are fed by any form of information that can be digitized: Only odors have yet to be transmitted by the information highways[9]. Messages circulate at speeds unheard of before in history. Time and distance are no longer limiting factors in the dissemination of knowledge and know-how. As soon as it is produced, an item of knowledge is delocalized, and reactions to it are almost as immediate, with the result that knowledge is constantly being reconfigured. The corollary is that the production system is equally modified. Individual intelligence, interfacing with that of a multitude of other thinking individuals, has become a

---

[9] In the near future the chemical formulas of odors will be digitized and the data automatically reconverted by a dissemination system on reception, as is already the case for sound.

distributed form of intelligence: Data processing takes place in parallel almost everywhere on the network, with output looping to input. Computer technology highlights the distributed, collective nature of intelligence; ultimately it is no longer the individual who thinks, but a society of interconnected minds.

— *"Who thinks? We no longer have one subject, or one thinking, 'material' or 'spiritual' substance. Thought is a network in which neurons, cognitive modules, human beings, teaching establishments, languages, writing systems, books and computers interconnect, transform and translate representations."* (Lévy, 1990, p. 156).

AI has enabled us to realize that human intelligence is both the source and product of ecology of the mind. No matter how brilliant, the mind cannot deny its dependence on the cultural and technical community from which it stems, and more than the body can survive without an exchange of energy and material with the environment. Just as the organism of which it is part, the mind organizes itself based on internal logic, but is fed from the outside. Autonomy and heterogenesis are not mutually exclusive: They are two possible ways of qualifying the operation of a system, corresponding to two points of view (inside/outside) or to two levels of description (formal/material). This distinction between the levels of observation is equally valid for computer networks: Each node in the network is materially formed by human mind-bodies linked to von Neumann computers, controlled top down. Locally, the system conforms to the computo-symbolic paradigm, but globally behaves as a neuronal network, with no central control: In networks such as Internet or multi-agent systems, macro-cognition seems to be governed by the connectionist paradigm, contrary to the postulate for the human mind. Must we therefore conclude that AI gives us a back-to-front image of reality, like any other mirror?! AI has confirmed the logical discontinuity from the individual to the collective, but above all has rendered perceptible the feedback of the global system to its constituent parts.

Quite apart from its technical prowess, AI has brought closer together object and subject, the sensible and the intelligible, reflection and action, the individual and the collective. These are not the least of its merits. In so doing, it has helped us to distance ourselves from the linear and dichotonomous ways of thinking that characterize modernity.

4   Conclusion

The computer revolution is twinned with a cultural revolution, whose implications have yet to be clearly perceived. "Think global, act local" is one of the leitmotifs of western culture in the second half of the 20th century. Telecommunications have heavily contributed to making that slogan widespread in the societal mind-set. Step by step, in connecting local subject-objects, AI has not only instigated planetary thought, but has also made possible collective, concerted action, for better or for worse. The possibility of acting on a large scale and from a distance does raise a certain number of truly new questions. The following are some of them:

- What will be the long-term effect of blurring the distinction between the real and virtual worlds on political decisions or on the individual psyche?
- Now that they can be manipulated at will, photographs can no longer provide supporting evidence for verbal testimony: "Pictures have lost their innocence" as Quéau has pointed out. In legal terms, will not the status of evidence such as written or photographic exhibits be considerably weakened?
- The notion of responsibility becomes problematic, since AI now seconds human beings in their decisions: If the diagnosis of an expert system turns out to be incorrect, or a computerized security system is defective, who is responsible? The program designer, the computer manufacturer, or the end-user?
- Faced with collective intelligence, how can we settle copyright issues, and more generally, that of intellectual property rights? This problem is not new, but has been amplified by the accessibility and speed of data transmission.
- Some believe that the accelerated dissemination of opinions and ideas can enhance democracy, since it is more difficult to control a network comprising several hundreds of thousands of servers distributed over five continents than several dozen radio-TV broadcasting stations and a few national press agencies.

- *"Far from being a means of control, on the contrary Internet will be a means to liberty, by enabling modern man to break free of bureaucracy. When information transcends national frontiers, States have much greater difficulty in governing by lies and propaganda. This can already be seen in computerized companies: Communication breaks out of the hierarchical channels, which thereby slowly flatten, with the result that anxious deference and arrogant certainty give way to dialogue on an egalitarian basis."* (Huitéma, 1996, p. 181).[10]

No technology ever had that kind of power in the past. Is AI safe from misuse? Probably not. It is no less true, however, that the networking of computers creates communities, no doubt virtually-based, but whose effects are quite real. The exchange of symbols and significations, images and ideas, stimulates co-operation and reflection, opening up new ways for human beings to see and be, which significantly change social practices.

Curiously enough, the cybernaut community has a certain number of characteristics in common with churches: Universality, user-friendliness, dedicated servers (cf. priests and parsons), the non-exclusion of persons (cf. women, foreigners or pagans). In its own way, Internet links up highly diverse beings

---

[10] Since 1991 Christian Huitéma has been a member of the IAB (Internet Architecture Board), which supervises the development of the Internet network.

that would otherwise be totally separated by their geographical remoteness, their nationality, education or social condition. Is cyberspace setting up a new form of transcendental institution, different from that of the State, the Red Cross, economics or science? And if such is the case, how can we qualify that transcendence? One may only see computerized collective intelligence as the extension to the ideal put forward by the Encyclopaedists: Making access to knowledge more democratic and enhancing the formation of opinion. Today, however, we ply, play, plan, and produce with the aid of computers. In its networks AI captures a good deal of human experience. As a planetary workshop of the imagination, Internet is a gigantic metaphor of the world. Not its duplicate, like Biosphere II[11], but a parable reflecting mankind. At the dawn of the third millennium, AI is a challenge for human intelligence to decipher what is going on with the virtual revolution.

References

Agre, P.E. (1988): *The dynamic structure of everyday life*, MIT Technical Report 1085 (Cambridge MA)

Amit, D. (1989): *Modeling Brain Function. The World of Attractor Neural Networks* (Cambridge University Press, Cambridge MA)

Andler, D. (1992): "From paleo to neo-connectionism", in: *New Perspectives on Cybernetics*, ed. by J. van der Vijver (Kluwer, Dordrecht), pp. 125–146

Atlan, H. (1986): *A tort et à raison. Intercritique de la science et du mythe* (Seuil, Paris)

Bateson, G. (1972): *Steps to an Ecology of Mind* (Ballantine Books, New York)

Bourret, P., Reggia, J. and Samuelides, M. (1991): *Réseaux neuronaux. Une approche connexionniste de l'intelligence artificielle* (Teknea, Toulouse / Marseille / Barcelone)

Clark, A. (1989): *Microcognition: Philosophy, Cognitive Science and Parallel Distributed Processing* (MIT Press, Cambridge MA)

Davis, R., Lenat, D.B. (1982): *Knowledge-Based Systems in Artificial Intelligence* (McGraw-Hill, New York)

Dobzhansky, Th. (1966): *L'homme en évolution* (Flammarion, Paris)

Ducret, J.-J. (1991): "Constructivisme génétique, cybernétique et intelligence artificielle", *Cahiers de la Fondation Archives Jean Piaget* 11, 19–40

Ducret, J.-J. (1991): "L'intelligence peut-elle être artificielle? Epistémologie génétique et intelligence artificielle, vecteurs d'une nouvelle conception de l'homme", *Le Nouveau Golem* 1, 13–32

Dupuy, J.-P. (1994): *Aux origines des sciences cognitives* (La Découverte, Paris)

Edelman, G. (1992): *Bright Air, Brillant Fire: on the Matter of the Mind* (Allen Lane, The Penguin Press, London)

---

[11] *Biosphere II* is a scientific program located in the Arizona desert which consists in creating a totally independent ecosystem comprising more than 2000 plant and animal species.

Faihe, Y., Müller, J.-P. and Rodriguez. M. (1996): "Designing Behaviour-based Learning Robots", submitted to IROS

Gaussier, Ph. and Zrehen, S. (1994): "A Topological Map for On-Line Learning: Emergence of Obstacle Avoidance in a Mobile Robot", it Proceedings of From Animals to Animats (MIT Press, Cambridge MA)

Grimson, W. E. L. and Patil, R. S. (1987): *AI in the 1980s and beyond* (MIT Press, Cambridge MA), pp. 343–363

Haase, K. (1984): *ARLO: The Implementation of a Language for Describing Representation Languages*, Bachelor's Thesis, Department of Philosophy and Linguistics, MIT

Haugeland, J. (1985): *Artificial intelligence, the very idea* (MIT Press, Cambridge MA), pp. 60–65

Heims, S. (1980): *John von Neumann and Norbert Wiener. From Mathematics to the Technologies of Life and Death* (MIT Press, Cambridge MA)

Heims, S. (1991): *The Cybernetics Group* (MIT Press, Cambridge MA)

Huitéma, Ch. (1995): *Et Dieu créa l'Internet* (Eyrolles, Paris)

Le Moigne, J.-L. (1995): *Les épistémologies constructivistes* (Presses Universitaires de France, Paris)

Lévy, P. (1990): *Les technologies de l'intelligence. L'avenir de la pensée à l'ère informatique* (La Découverte, Paris)

Lévy, P. (1995): *Qu'est-ce que le virtuel?* (La Découverte, Paris)

Mallery, J.C., Hurwitz, R., and Duffy, G. (1986): "Hermeneutics: From textual Explication to Computer Understanding?". MIT, AI Memo 871, 1-32, published in: *The Encyclopedia of Artificial Intelligence*, 1987, edited by S.C. Shapiro (Wiley, New York)

Memmi, D. (1990): "Connexionnisme, intelligence artificielle et modélisation cognitive", *Intellectica* 9/10

Minsky, M. (1975): "A Framework for Representing Knowledge", in: *The Psychology of Computer Vision* (McGraw-Hill, New York), pp. 211–277

Minsky, M. (1986): *The Society of Mind* (Simon and Schuster, New York)

Müller, J.-P. and Rodriguez, M. (1995): "Representation and Planning for Behavior-based Robot", in: *Environment Modeling and Motion Planning for Autonomous Robots*, edited by H. Bunke and T. Kanade (World Scientific, Singapore)

Müller, J.-P. and Pecchiari, P. (1996): "Un modèle de systèmes d'agents autonomes situés: application à la déduction automatique", in: *IA distribuée et systèmes multi-agents*, ed. by J.-P. Müller and J. Quiqueton (Hermès, Paris)

Penrose, R. (1994): *Shadows of the Mind* (Oxford University Press, Oxford)

Piaget, J. (1937): *La construction du réel chez l'enfant* (Delachaux et Niestlé, Neuchâtel)

Proust, J. (1987): "L'intelligence artificielle comme philosophie", *Le Débat* 47, 88–102

Quéau, Ph. 1993): *Le virtuel. Vertus et vertiges* (Champ Vallon, Seyssel)

Quillian, R. (1967): "Semantic Memory", in: *Semantic Information Processing*, (MIT Press, Cambridge MA), pp. 227–270

Reeke, G. and Edelman, G.R. (1989): "Real Brains and Artificial Intelligence", in: *The Artificial Intelligence Debate*, ed. by S. Graubard (MIT Press, Cambridge MA)

Ricoeur, P. (1986): *Du texte à l'action* (Seuil, Paris)

Rorty, R. (1981): *Philosophy and the Mirror of Nature* (Princeton University Press, Princeton)

Rumelhart, D.E. and McClelland, J.-L. (1986): *Parallel distributed Processing. Explorations in the Microstructure of Cognition* (MIT Press, Cambridge MA)

Schank, R and Abelson, R. (1977): *Scripts, Plans, Goals and Understanding* (Erlbaum, Hillsdale N.J.)

Searle, J. (1980): "Minds, Brains and Programs", *Behavioral and Brain Sciences* 3, 417–458

Smolensky, P. (1988): "On the Proper Treatment of Connectionism", *Behavior and Brain Sciences* 11,1–74

Steiner, E.H., Schmidt, P. and Zelinsky-Wibbelt, C. (1988): *From Syntax to Semantics; Insights from Machine Translation* (Pinter, London)

Stengers, I. (1987): *D'une science à l'autre. Des concepts nomades* (Seuil, Paris)

Turing, A. (1937): "Computability and Lambda-definability", *Journal of Symbolic Logic* 2

Van Gelder, T. and Port, R. (1994): *Mind as Motion* (MIT Press, Cambridge MA)

Varela, F.J. (1996): *Invitation aux sciences cognitives* (Seuil, Paris), (revised edition of: *Cognitive Science. A Cartography of Current Ideas, 1988*)

Von Neumann, J. (1966): *Theory of Self-Reproducing Automata* (University of Illinois Press, Chicago)

Waltz, D.L. (1989): *Semantic Structures, Advances in Natural Language Processing* (Erlbaum, Hillsdale N.J.)

Zrehen, S. (1995): *Elements of Brain Design for Autonomous Agents*, Bachelor's Thesis, EPFL, Department of Informatics (Lausanne, Switzerland)

Self-consciousness as a Philosophical Problem

Thomas Spitzley

FB Philosophie, Gerhard-Mercator-Universität Gesamthochschule Duisburg,
D-47048 Duisburg, Germany. e-mail: spitzley@uni-duisburg.de

The title "Self-consciousness as a Philosophical Problem" is ambiguous. On
the one hand, it can be seen as what philosophers understand as "self-
consciousness". On the other hand, one could ask if that with which philosophers have dealt under this heading indeed were philosophical problems because it may well be that they are nothing but (disguised) empirical questions.
In this chapter I shall try to present various problems related to this topic
so that they will become recognizable as just different facets of one and the
same problem.

After a short excursion into the history of the concepts of consciousness,
self, and self-consciousness I shall delve into the search for the self, whereby I
shall follow (more or less) the development in the history of philosophy. The
search for the self will begin under the label "What is it that the self is conscious of?". After dealing with Descartes' discovery of the self, I shall discuss
how the idea that the self has a consciousness could be characterized more
precisely. Then, asking "Do we have a consciousness of our self?", our path
will lead us past Hume and on to Kant, focusing on the nature of the self. The
final section can be characterized by the slogan "One can do without a self,
or: from one's self to oneself". There I shall discuss the considerations of the
modern philosophy of language which can be seen as attempts at explaining
what it means to have a consciousness of oneself.

1 On the History of "Consciousness", "Self", and "Self-consciousness"

The first recorded occurrences of the expression "consciousness" as well as
of the word "self-consciousness" can be found in the 17th century. Their
philosophical usage was established by John Locke. The English word "consciousness" most likely goes back to Descartes' concept of *conscientia*. In
scholasticism this term was used mainly in the sense of "conscience",[1] yet
in Descartes *conscientia* is always found in connection with the concept of
cogitatio. The *con-scientia* is something like a 'with-knowledge'; it is knowl-

---

[1] As to the close connection between conscience and consciousness cf. "**Conscience**
[...] The consciousness of moral worth or its opposite as manifested in character
or conduct, together with the consciousness of personal obligation to act in accordance with morality and the consciousness of merit or guilt in acting." (Baldwin
(Ed.), *Dictionary of Philosophy and Psychology*, Vol. 1, p. 215).

edge of the fact that it is we who perform certain acts.[2] So the concept of consciousness is closely related to a quasi-internalized confidant or observer of our acts and thoughts.

In its substantival use, the expression "self" can be traced back to at least the 13th century. The *Oxford English Dictionary* has under the entry "self", e.g.:

> "That which a person is really and intrinsically *he* (in contradistinction to what is adventitious); the ego (often identified with the soul or mind as opposed to the body); a permanent subject of successive and varying states of consciousness.
>
> *a* 1674 TRAHERNE *Poet. Wks.* (1903) 49 A secret self I had enclos'd within, That was not bound with my clothes or skin. 1682 SIR T. BROWNE *Chr. Mor.* 1. §24 The noblest Digladiation is in the Theater of our selves. [...]
>
> What one is at a particular time or in a particular aspect or relation; one's nature, character, or (sometimes) physical constitution or appearance [...]
>
> An assemblage of characteristics and dispositions which may be conceived as constituting one of various conflicting personalities within a human being. *better self:* the better part of one's nature.
>
> [...] *a* 1703 BURKITT *On N.T.* Mark xii. 34 Every man may, yea, ought to love himself; not his sinful self, but his natural self: especially his spiritual self, the new nature in him."[3]

In its colloquial usage "self-consciousness" is an evaluative term. In general when someone is called "self-conscious", it is to be understood negatively. A self-conscious person is "[m]arked by undue or morbid preoccupation with one's own personality; so far self-centred as to suppose one is the object of observation by others".[4]

Interestingly enough, the expression that is used in German philosophical discourse about self-consciousness has a positive connotation: to call someone "selbstbewußt" means to attribute to him a conviction of his abilities and of his personal value as well as importance.[5]

This much is sure, "self-consciousness" as a technical philosophical term is neither negatively nor positively connotated. If one wants to come closer to the philosophical meaning of "self-consciousness", another term, which is well-known from the history of philosophy, can help, viz. "self-knowledge". He who wants to follow the delphic oracular utterance *gnothi se auton* ("know thyself"), will find out something about himself, his abilities, his true wishes and beliefs, instead of yielding perhaps to self-deception. So self-knowledge

---

[2] Cf. Ritter (Ed.), *Historisches Wörterbuch der Philosophie*, Vol. 1, p. 891.

[3] *The Oxford English Dictionary,* Vol. 14, pp. 906 f.

[4] *The Oxford English Dictionary,* Vol. 14, p. 916.

[5] Cf. *Duden. Das große Wörterbuch der deutschen Sprache,* Vol. 5, p. 2374.

has an immediate practical benefit. It is precisely the aspect, that you *learn* something about yourself during the process of acquiring self-knowledge, which points to the difference between self-knowledge and self-consciousness (in its philosophical meaning): contrary to this kind of self-knowledge which may be difficult and strenous to acquire, the self-consciousness that philosophers are talking about is said to be something we simply possess.

2   What Is It That the Self Is Conscious of?

Descartes is considered to be the father of the philosophical theory of self-consciousness.[6] In his *Meditations on First Philosophy*, which were published in 1641, he tries to establish, with the assistance of pure reason only, by means of methodic and radical doubt, what we can safely rely upon and what we can hold to be true with certainty.

The first of four rules which Descartes proposes for the use of reason is "to accept nothing as true which I did not clearly recognize to be so: that is to say, carefully to avoid precipitation and prejudice in judgments, and to accept in them nothing more than what was presented to my mind so clearly and distinctly that I could have no occasion to doubt it."[7] If one follows this rule, one can hope to eventually reach a *fundamentum inconcussum*, an unshakeable, secure foundation upon which the entire structure of our knowledge can be established.

Using this rule, Descartes first of all doubts the perceptions of his senses. To be sure, our senses deceive us once in a while, but aren't there things which can be traced back to sensory perceptions and of which we can be certain, like, as Descartes writes, "that I am here, seated by the fire, attired in a dressing gown, having this paper in my hands and other similar matters"?[8] It seems that we must doubt all our sensory perceptions because we could be sleeping and experiencing everything in a dream. Even if that is true, it still seems that there are some things which are exempt from doubt: "For whether I am awake or asleep", Descartes says, "two and three together always form five, and the square can never have more than four sides".[9] Are, therefore, arithmetic and geometry two areas in which we can achieve certainty? No, in the end they are not, since God could make it so that we constantly err when we add or when we count the sides of a square.

Here is where the radical element of the methodic doubt comes into play: Descartes suggests that we should doubt *everything*, or even "for a certain time" hold all opinions to be "entirely false and imaginary"[10] and suppose

---

[6] A complete history of the philosophical term "self-consciousness" though, would have to begin with Augustin, if not earlier (cf. Scholz, Augustin und Descartes).

[7] Descartes, Discourse on the Method of Rightly Conducting the Reason, p. 92.

[8] Descartes, Meditations on First Philosophy, p. 145.

[9] Descartes, Meditations on First Philosophy, p. 147.

[10] Descartes, Meditations on First Philosophy, p. 148.

that a *genius malignus* is trying to deceive us in every possible way. If one accepts that there is an all-powerful deceiver, nothing can be accepted as true which could have been influenced by such a deceiver. Yet an evil demon may deceive me as much as he likes, "he can never cause me to be nothing so long as I think that I am something".[11] This wording is a variation of Descartes' famous Cogito-argument "I think, therefore I am".[12]

Certain, according to Descartes, is that I exist when and as long as I think. It is still unclear, however, who or what I am. We are inclined to answer unbiased "a human being", but this possible answer is denied us because every human being has a body and an evil spirit could make me believe, for example, that I possess hands, even if it were not the case. If I can doubt that I have a body, I can also doubt that I have sensations because sensations presuppose a body.[13] Alone my thinking, my consciousness (*cogitatio*) escapes doubt: "[...] it alone cannot be separated from me. I am, I exist, that is certain".[14] For Descartes the result of this is that I am nothing but a *res cogitans*, a thinking thing, i.e. "a thing which doubts, understands, [conceives], affirms, denies, wills, refuses, which also imagines and feels"[15] – which, by the way, indicates that the word "cogitare" has a much broader meaning than does the English word "think". So I am a *res cogitans*; my body, however, and all the other things in the so-called outer world belong to the area of *res extensae*, the extended things. The *res cogitans*, the I or the self, is supposed to be an immaterial substance which is not extended, which possesses no parts, which consists of itself and "whose whole essence or nature is to think".[16]

According to Descartes' explanation of self-consiousness, the self-certitude of the thinking I is the Archimedean point from which one can reject all the sceptical considerations. Descartes tries to show how this is possible during the course of his *Meditations*, although I shall not elaborate farther on this

---

[11] Descartes, Meditations on First Philosophy, p. 150.

[12] Carriero points out that the Cogito-argument has some predecessors, viz. Thomas Aquinas, *De Veritate*, Q. 10, A. 12, ad 7: "in thinking something, he perceives that he exists", and Aristotle, *Nicomachian Ethics*, 1170a30: "if we perceive, we perceive that we perceive, and if we think, that we think; [...] ([...] existence was defined as perceiving or thinking)". Cf. Carriero, The Second Meditation and the Essence of Mind, esp. p. 202 and p. 208.

[13] Descartes, Meditations on First Philosophy, p. 151.

[14] Descartes, Meditations on First Philosophy, p. 151.

[15] Descartes, Meditations on First Philosophy, p. 153. – With regard to the last element some caution is necessary. Descartes had maintained that having sensations presupposes a body (cf. p. 151); when he now claimes that a *res cogitans* be a thing which feels, he should more precisely claim it to be a thing which (among other things) *seems* to feel (cf. p. 153).

[16] Descartes, Meditations on First Philosophy, p. 190; cf. p. 165 and p. 196. – "[O]ne *cannot* (according to Cartesian doctrine) observe substances directly. They must be known through their attributes." (Wilson, *Descartes*, p. 66; cf. Descartes, The Principles of Philosophy, p. 240)

here. According to Descartes, the consciousness which the self, the I, has of itself as a thinking being is the foundation upon which all our knowledge is based. This consciousness of the self is (in the end) nothing other than the knowledge that the I or the self has with respect to its states of consciousness or, as one can also say, with respect to its *psychological states*,[17] as long as the self is in these states.

When Descartes says "I am a thinking thing", it seems to indicate that one should differentiate between a thing which thinks and the individual acts of thinking. On the one hand there seem to be the various *psychological states*, and on the other hand there are the *bearers* of these psychological states. Here self-consciousness is to be understood as a relation whose relata are psychological states and a self. So it becomes clear once again that in this sense self-consciousness is to be explicated as the consciousness, the knowledge which a self has of its psychological states.

The following discussion will focus on both of these relata, the psychological states and the self. I shall consider which kind of access to our psychological states we have, which characteristics our judgements concerning our present states of consciousness possess, what the self could be and how we can possibly gain access to it. In this discussion the term "judgement" should be understood, following Frege, as "thought which has been acknowledged as true".[18] If I judge that Chirac is the President of France, then I acknowledge the thought that Chirac is the President of France as true.

At the end of the linguistic-historical part I had said that self-consciousness is something that we are all endowed with and that it is something which we do not have to acquire. Then I showed that one can understand self-consciousness as the consciousness that a self possesses. Descartes' position amounts to the thesis that self-consciousness is a knowledge which a self – or rather: my self, my 'I' – has of its own present psychological states.

A problem seems to arise here, because don't we, to some extent, first have to acquire the knowledge we have of our psychological states? Did we not learn, at the latest from Freud, that we are not always aware of all of our wishes, intentions, opinions, and feelings, and that it sometimes takes going through a painful process before we can free ourselves from a false self-image? Yes, this is all true, but it only makes apparent that today we cannot adhere to certain theses which may rightly be attributed to Descartes, at least not with their original scope.

There are some reasons to assume that Descartes thought that the judgements which we make about our own present psychological states are infallible, and that these psychological states are epistemologically transparent.

---

[17] For the sake of simplicity, I only talk about psychological states, although, if one wants to be more precise, not only psychological *states* are at issue here but also psychological *acts*. With regard to psychological states one can (roughly) distinguish between propositional attitudes and phenomenological states.

[18] Cf. Frege, Thoughts, p. 7.

If this is true, we can deceive ourselves neither regarding their presence or absence nor with respect to their characteristics.[19]

Descartes' position, which was just described, can be explained in the following way: (a) If I *believe* that I am in a certain psychological state, then it is *true* that I am in that psychological state, i.e., I cannot err with regard to my being in a certain psychological state. To put it another way, my judgements regarding this are infallible. (b) If it is *true* that I am in a certain psycholological state, then I *believe* that I am in this state, and if it is *not* true that I am in a certain psychological state, then I *do not believe* that I am in that state. That is to say, my psychological states are epistemologically transparent. In other words, I am omniscient with respect to the psychological states which I am in. The discussion about whether our judgements concerning our own present mental states[20] have special epistemological characteristics, can be summarized under the heading "First Person Authority". This title expresses the idea that a person who is in a certain psychological state has a special authority with regard to whether it is true that he is in a certain psychological state or not.

Both of these theses from Descartes, that I am infallible in my judgements about my present psychological states and that all these states are epistemologically transparent for me, are still, even today, supported by many philosophers, perhaps not with this intensity, but at least in a milder variation. *One form of modification is to simply change their scope.* It has been claimed, for example, that infallibility did not hold true for *all* our judgements about our present psychological states, but it was indeed at least valid for a subclass of these judgements, viz., e.g., for our judgements about some of our sensations, and these sensations were epistemologically transparent as well. If I believe that I am in pain, then it is also true that I am in pain. Likewise, if I am in pain, then I also know that I am in pain; if I am not in pain, then I also do not believe that I am in pain; and if I do not believe that I am in pain, then I am not in pain.

To what extent one possesses an authority concerning judgements of the form "I am now in such-and-such a psychological state", and how far this authority reaches is quite controversial. Most philosophers only agree to the fact that there is an *epistemological asymmetry*, as Tugendhat calls it, between judgements about one's own present psychological states (formulated in the first person singular: "I am now in such-and-such a psychological state") and judgements concerning the present psychological states of another person.[21]

It seems that different epistemological predicates apply to *my* judgements about my present psychological states than to *another person's* judgements about my present psychological states. *Another person* can, as is traditionally claimed, only conjecture that I am in pain or only believe that I am of this

---

[19] Cf. Wilson, *Descartes*, pp. 151 ff.

[20] I use the expressions "state of consciousness" and "mental state" as synonyms.

[21] Tugendhat, *Selbstbewußtsein und Selbstbestimmung*, p. 89.

or that opinion. *I*, however, know if I am in pain, and I know what I believe; I am sure of these things and cannot err in my judgements.

Although there seems to be an epistemological *a*symmetry between first and third person judgements, there is also something they have in common, for both of them always have the same truth-value: "The sentence 'I am now in such-and-such a psychological state', if it is expressed by me, is true if and only if the sentence 'he is now in such-and-such a psychological state', if it is uttered by someone else who means me when he says 'he', is true."[22] Tugendhat calls this characteristic *"veritative symmetry"*.[23]

Apart from the intense discussion about the *kind* of authority that the first person has, this authority has also been criticized in principle. Firstly, there have been objections to the thesis of epistemological asymmetry. According to these objections, the first person is said to be in no way epistemologically privileged even if he does sometimes in some ways possess advantages over the third person. I shall come back to this point later.[24]

Secondly, the thesis of veritative symmetry has been attacked. This, at least at first sight, very surprising objection, which can be traced back to Wittgenstein, does not state that "He is now in such-and-such a psychological state" can be false, even if "I am now in such-and-such a psychological state" is true (or vice versa). Instead it aims at denying that the statement "I am now in such-and-such psychological state" has any truth-value at all – at least in certain contexts.

For Wittgenstein, "I am in pain" is the prime example of an expression of a sensation. Such an avowal is a learned behavior for pain which often replaces a scream which would be a 'natural expression of the sensation'.[25] According to Wittgenstein, such sentences are not always uniformly used, instead one can make an exclamation or a confession with them, or can give an explanation or can, e.g., report something to a doctor. In his opinion, however, the expressive use is the primary use. If, though, "I am in pain" is only a cry of complaint,[26] then this avowal can be sincere or insincere, but it does not have a truth-value, just as a scream does not have a truth-value.

According to Wittgenstein, statements about one's own present psychological states do not only have the above-mentioned expressive character, but also a non-cognitive character. This means that a speaker does not express a

---

[22] "Der Satz 'ich befinde mich gerade in dem-und-dem psychischen Zustand', wenn er von mir geäußert wird, ist wahr genau dann, wenn der Satz 'er befindet sich gerade in dem-und-dem psychischen Zustand', wenn er von jemand anderem geäußert wird, der mit 'er' mich meint, wahr ist." (Tugendhat, *Selbstbewußtsein und Selbstbestimmung*, p. 88; my translation into English) In this quotation I have substituted "befinde(t) mich (sich) gerade in dem-und-dem psychischen Zustand" for "*Φ*", which Tugendhat uses as a variable for psychological predicates.

[23] Cf. Tugendhat, *Selbstbewußtsein und Selbstbestimmung*, p. 89.

[24] Cf. below, p. 49.

[25] Cf. Wittgenstein, *Philosophical Investigations*, §§ 244 f, 256, 304, 291, 582, 585.

[26] Cf. Wittgenstein, *Philosophical Investigations*, p. 189.

claim to knowledge with the statement "I am in pain."[27] During one phase of his work Wittgenstein supported the idea that "I know that I am in pain" at best means the same as "I am in pain";[28] later he at least admitted that "I know that I am in pain" could be used as an emphatic expression of pain. In addition, there is only one other possible use for sentences like "I know that I am in pain" or "Only I can know what I feel or think", namely their use as *grammatical sentences*. For Wittgenstein "grammatical sentence" is a technical term which must not be confused with "sentence out of a grammar book". Grammatical sentences express rules, or at least partial rules, for the use of individual words, i.e, rules which, as it were, describe the bounds of sense, the boundaries of meaningful speech. "In this grammatical use they [i.e., sentences like "I know that I am in pain" or "Only I can know what I feel or think"] state among other things that in the first person doubt and uncertainty are meaningless, that a speaker's sincere utterances have an authoritative status, that he can hide his thoughts if he wants to, etc."[29]

Wittgenstein writes on this topic:

"I can know what someone else is thinking, not what I am thinking.
It is correct to say 'I know what you are thinking', and wrong to say 'I know what I am thinking.' (A whole cloud of philosophy condensed into a drop of grammar.)"[30]

In this quote Wittgenstein reveals himself as an advocate of a completely different kind of asymmetry: we can indeed know what *someone else* thinks, but not what *we ourselves* think; we can know that *someone else* is in pain, but not that *we* are in pain.

The decisive reason for this is that one can meaningfully ask how one can find out if *someone else* is in pain, but that it would be senseless to ask how one could find out if *oneself* is in pain. In order to be able to determine whether we ourselves are in pain we do not need any kind of criterion. Yet such criteria and the possibility to convince oneself, or to doubt, just to suppose or to err that one is in a certain state are necessary if it is to make sense to talk about knowledge.[31]

---

[27] Cf. Wittgenstein, *Philosophical Investigations*, p. 221.

[28] Wittgenstein, The Blue Book, p. 51, and Wittgenstein, Notes for Lectures on 'Private Experience' and 'Sense-Data', p. 309.

[29] "In dieser grammatischen Verwendung legen sie [sc. die Sätze "Ich weiß, daß ich Schmerzen habe" und "Nur ich kann wissen, was ich fühle oder denke"] unter anderem fest, daß in der ersten Person Zweifel oder Unsicherheit sinnlos sind, daß die aufrichtigen Äußerungen des Sprechers einen autoritativen Status haben, daß dieser seine Gedanken verbergen kann, wenn er will, etc." (Glock, Innen und Außen: "Eine ganze Wolke Philosophie kondensiert zu einem Tröpfchen Sprachlehre", p. 243; my translation into English)

[30] Wittgenstein, *Philosophical Investigations*, p. 222.

[31] Cf. Wittgenstein, *Philosophical Investigations*, p. 221.

This will have to be enough about Wittgenstein's criticism of the thesis of *veritative symmetry*, the thesis according to which first person judgements about psychological states and the corresponding third person judgements always have the same truth-value. I shall now come to the fundamental opposition to the *epistemological asymmetry* between these two types of judgements.

Is a person who self-ascribes psychological states really in an epistemologically privileged situation over a person who ascribes psychological states to someone else? Traditionally, the answer to this question has been "yes", and for the following reason: what would I have to do if I wanted to find out if a certain person is now in pain or would now like to drink a cup of coffee? I have two possibilities: I can observe his behavior (e.g. his facial expression), and I can ask him – hoping to get an honest answer. My observations or his answers would give me the criteria by means of which I could decide what psychological state he is in. *My* access to *his* psychological states is, therefore, only indirect, mediated, for I need evidence in order to come to my conclusion, and of course I could make a mistake.

Now what would I have to do if I wanted to find out if *I* am in pain or would like to drink a cup of coffee? Well, nothing – in a way – since contrary to ascribing psychological states to another person most philosophers agree that self-ascriptions are made without the help of any criteria. So, I have direct access to my present psychological states; these states are 'given' to me directly; that I am now in a certain state is evident to me and introspection – be that what it may – imparts to me the respective knowledge, and this knowledge can be neither checked by a third person nor can it be subject to the possibility of an error.

So a person who ascribes psychological states to himself is indeed in an epistemologically privileged situation over a person who ascribes psychological states to another? An emphatic "No!" is the answer from Gilbert Ryle, who created furore with his work *The Concept of the Mind* some 40 years ago. In this book he argued vehemently against the cartesian dualism of body and mind which he called 'the dogma of the Ghost in the Machine'.[32] Ryle supports a definite behavioristic position, according to which psychological states are not immaterial, but must simply be understood as dispositions of behavior. In his opinion, the methods by which I find something out about myself are basically the same as those by which I can find something out about someone else.[33] Our knowledge of others as well as ourselves depends on our observing their behavior or ours, respectively.[34] If we want to acquire self-consciousness, we have to take notice of our own unreflected, spontaneous utterances, "including our explicit avowals, whether these are spoken aloud, muttered, or said in our heads".[35] Therefore, Ryle claims, self-consciousness

---

[32] Ryle, *The Concept of Mind*, p. 17.
[33] Ryle, *The Concept of Mind*, p. 149.
[34] Ryle, *The Concept of Mind*, p. 173.
[35] Ryle, *The Concept of Mind*, p. 176.

"is simply a special case of an ordinary more or less efficient handling of a less or more honest and intelligent witness."[36] If one accepts Ryle's theory, there is no longer any *epistemological asymmetry* between judgements about one's own present psychological states and judgements about the psychological states in which someone else is now in. Ryle concedes, however, that *in some respect* every person is indeed privileged regarding his own mental states, viz. in that generally one does possess the most observational data of one's own behavior. This is, however, only a practical advantage and not one in principle.

With this I should like to conclude the discussion of the question what it is that the self is conscious of. I shall now address the problem of whether we have access to the putative bearer of the various psychological states, the self, and if so, what kind of access it is. So "self-consciousness" – meaning "a self which has a consciousness of its psychological states" – is to be understood differently now, namely as "being conscious of one's self".

3   Do We Have a Consciousness of Our Self?

It used to be presumed that we do have a consciousness of our self, and that the crucial ability for having this is introspection. Fairly unclear, however, is what is to be understood by introspection. It is natural to assume that it is a special kind of perception, namely, as the word's literal meaning suggests, a looking-into-oneself by which one possesses direct and unimpeded access to one's attitudes and sensations. Accordingly, introspection is something like an inner sense which is supplemented by the five outer senses: hearing, seeing, smelling, tasting, and feeling.

Two characteristics of introspection are obvious: (1) there is no special organ which makes introspection possible; (2) introspective perception is limited to the consciousness of the person who is 'introspecting'. In the case of genuine perception "[t]here is such a thing as singling out one [object] from a multiplicity of perceived objects, distinguishing it from the others (which may be of the same kind as it) by its perceived properties and its position in a space of perceived objects".[37] A perceived object can be identified as being of a certain kind, e.g., human or animal – and with this *sortal* identification mistakes can be made. One can also identify it as a certain object within a certain kind, e.g., as Ella F. or as Louis A. – and with this *particular* identification mistakes are possible, too. Finally, one can perceive an object at a time t_0, and then, at a later time t_1, perceive it again and identify it as the object which was perceived at t_0 – and with this *re*-identification,[38] mistakes are, of course, possible, too.

---

[36] Ryle, *The Concept of Mind*, p. 186.
[37] Shoemaker, Introspection and the Self, p. 108.
[38] Cf. Strawson, *Individuals*, pp. 32 ff.

None of these characteristics of perception seem to be involved when one is introspectively conscious of one's self: (1) Do we have to know any sort of object to which our self belongs? (2) Does it make sense to ask how we, if we do have a consciousness of ourself, distinguish our self from some other selves? (3) Are there really criteria with whose help we re-identify our self, i.e., make sure that it is still the same self that we refer to at two different points of time? All three questions are definitely to be answered "No"....

One of the most prominent critics of the view that there is a self in the sense that it is an independent, simple substance of which we also have direct, introspective knowledge is David Hume.

"Now everybody may think his own I and take notice of how he does it", asks Fichte of us.[39] Hume's famous and notorious, if anachronistic, answer can be found in his *Treatise of Human Nature*:

> "For my part, when I enter most intimately into what I call *myself*, I always stumble on some particular perception or other, of heat or cold, light or shade, love or hatred, pain or pleasure. I never catch *myself* at any time without a perception, and never can observe any thing but the perception. [...] If any one upon serious and unprejudic'd reflexion, thinks he has a different notion of *himself*, I must confess I can reason no longer with him."[40]

It is not surprising that Hume declares that he cannot find a self, no matter how hard he tries. In his opinion, theoretical reasons alone suffice for the fact that the kind of self he was looking for, an immaterial, simple substance, cannot be found in the characterized way. In order to be at all able to talk meaningfully about an immaterial substance, we would have to have an idea of this substance; since every idea was derived from a preceding impression and every impression was eventually based on perception,[41] we would have to be able to have a perception of the immaterial substance which the self is supposed to be,[42] and that is impossible.

Hume does indeed say that he is not able to find *this* kind of a self, but that, of course, does not mean that there is absolutely no self. Hume himself offers an alternative: the self is "nothing but a bundle or collection of different perceptions, which succeed each other with an inconceivable rapidity, and are in perceptual flux and movement."[43] Accordingly, the self is not something unchangeable, but rather something which is subject to constant changes. Naturally some questions remain unanswered: which feature of a perception, of a psychological state, is it that enables one to call this psychological state

---

[39] "Es denke nun jeder sein Ich, und gebe dabei Achtung wie, er es mache." (Fichte, *Wissenschaftslehre 1798 nova methodo*, p. 354)

[40] Hume, *A Treatise of Human Nature*, p. 252.

[41] Hume, *A Treatise of Human Nature*, p. 232 ff.

[42] Cf. Hume, *A Treatise of Human Nature*, p. 633.

[43] Hume, *A Treatise of Human Nature*, p. 252.

his psychological state? What is it that makes any set of psychological states into the delimitable, continually existing, and recognizable bundle which our self is supposed to be? Finally, who or what can gain knowledge of the individual psychological states or of the entire bundle if not the self – and thus this bundle of psychological states?

Hume failed in his attempt to find by way of experience that which Descartes believed he had found by means of reason, viz. the self. So the question of what this self is of which we are supposed to have a consciousness seems still to be open.

Kant's theory of self-consciousness is a critical reaction to Descartes as well as to Hume. That Hume, in his search of the self, only found psychological states can be due to what Kant called *"empirical* self-consciousness". To have empirical self-consciousness means to have knowledge of one's own states of consciousness. That was the point in the context of the discussion about privileged access and first person authority.

Kant has a ready answer to the question of which self is meant in talking about *empirical* self-consciousness: the empirical subject of the experience is the person himself, i.e., the person who *has* the consciousness, so that we no longer need to hold on to Hume's bundle-theory of psychological states.

Yet how can it be explained that one can attribute various psychological states to oneself and know that it is always the same self to which one attributes these various states? What determines the affiliation between a psychological state and a certain self? Also this Humean problem, namely how to tie the bundle of psychological states together, is one that Kant believes he can solve. He claims the decisive criterion to be that it is "possible for the 'I think' to accompany all my representations".[44] It's that easy; whenever I can, with regard to any representation, say "I think: ..."(followed by an expression of a representation) then this representation is *my* representation.

This "'I think' [for which it must be possible] to accompany all my representations" serves yet another purpose for Kant and is then called *"transcendental* self-consciousness". Transcendental self-consciousness rests on an achievement of synthesis: the process of synthesis enables us to see our perceptions and representations as being combined with each other.[45] "I can count a given representation as *mine* solely because *I* have combined or synthesized it with others."[46]

The transcendental self-consciousness is not a special kind of experience which we can have of ourselves; instead it is the condition of the possibility of empirical self-ascription of experience. It is in this respect "that 'transcendental self-consciousness' is the core of empirical self-consciousness."[47]

---

[44] Kant, *Critique of Pure Reason,* B 131.
[45] Kant, *Critique of Pure Reason,* B 133, A 108 and B 134.
[46] Strawson, *The Bounds of Sense,* p. 94.
[47] Strawson, *The Bounds of Sense,* p. 111.

Incidentally, Kant was not interested in self-consciousness as a problem *sui generis*, but rather considered transcendental self-consciousness to be indispensable for defending his general theory of knowledge.

Transcendental self-consciousness is a consciousness of oneself as an object to which there is a certain continuity and unity. "I am conscious of the self as identical in respect of the manifold of representations that are given to me in an intuition, because I call them one and all *my representations*, which constitute *one* representation."[48] And this one representation *is* the identical self.

The representation which one has of one's self, one's transcendental I, is, according to Kant, "in itself completely empty".[49] Of this I, says Kant, it is absolutely impossible to know its nature.[50] In Kant's opinion even the question as to the nature of the self is ill-conceived, if one considers the self to be an entity, material or immaterial. This is – in a condensed form – the result of Kant's way of critically dealing with Descartes' thesis according to which the I is an immaterial and simple substance.

4 One Can Do Without a Self, or: From One's Self to Oneself

If the self is not such an entity, like the above-mentioned entity, then what is it? Couldn't it be that the suggestion made by the Australian philosopher David M. Armstrong leads us in the right direction? Armstrong maintains that the self, *the mind*, as he instead says, is "something that is *postulated* to link together all the individual happenings of which introspection makes us aware. In speaking of minds, perhaps even in using the word 'I' in the course of introspective reports, we go beyond what is introspectively observed. Ordinary language here embodies a certain theory."[51]

Even if one does not want to follow Armstrong's proposal to consider the self a theoretical construct, there is still a different method clearly recognizable in this quote, the application of which could prove to be fruitful.

---

[48] Kant, *Critique of Pure Reason,* B 135: "Ich bin mir also des identischen Selbst bewußt, in Ansehung des Mannigfaltigen der mir in einer Anschauung gegebenen Vorstellungen, weil ich sie insgesamt *meine Vorstellungen* nenne, die *eine* ausmachen." Here I deviate in two respects from Kemp Smith's translation: (1) as in the German original, I also emphasize "representations"; (2) I interpret Kant as talking in the last bit of the quotation about a *representation* and not about an *intuition,* and therefore I prefer "which constitute *one* representation" to Kemp Smith's translation "and so apprehend them as constituting *one* intuition".

[49] Kant, *Critique of Pure Reason,* B 404.

[50] Of this I "ist schlechterdings nichts weiter zu erkennen möglich, was es für ein Wesen, und von welcher Naturbeschaffenheit es sei [...] " (Kant, Welches sind die wirklichen Fortschritte, die die Metaphysik seit Leibnizens und Wolffs Zeiten in Deutschland gemacht hat?, A 36)

[51] Armstrong, *A Materialist Theory of Mind,* p. 337.

Another way to find out what could be meant by the ominous self is, namely, to examine more closely how we refer to it. Is it not always so that when someone is talking about *my 'self'*, he is talking about *me*? And do I not, as a rule, use the expression "I" when I speak of myself? Don't I ask then, when asking about my 'self', always about that which I refer to with the expression "I"? This should not, however, give the impression that the latter be easy to answer. One thing, however, is clear: the question as to the nature of the 'self' turns into a question concerning the use of, or better yet, the reference of the personal pronoun "I". So in this respect the question as to the nature of self-consciousness presents itself cloaked in terminology borrowed from the philosophy of language.[52]

He who inquires into the reference of the pronoun "I" is trying to find out for which object this word stands. The expression "morning star" stands for the planet Venus, e.g., and the proper name "Jacques Chirac" denotes the present President of France.

With regard to the discussion about the reference of "I", which has been especially controversial during the last 20 years, one can observe a phenomenon which is typical for philosophy: there is not even any agreement concerning whether "I" is a referring expression at all. The prevailing view is, however, that it *does* refer, although there is no consensus as *to what* it refers.

The Cambridge philosopher Elisabeth Anscombe supports a radical position. The central thesis of her article "The First Person" from 1975 reads: " 'I' is neither a name nor any other kind of expression whose logical role is to make reference, at all."[53] The word "I" belongs in the same boat with the pronoun "it," for example. Just as the "it" in "it is thundering" does not stand for any object which thunders (even if the ancient Greeks may have had a different opinion on this topic), – according to Anscombe – neither does the "I" in "I am in pain" stand for an object which is in pain. Her argument for this claim is basically a *reductio ad absurdum*: if "I" really stood for an object, then this object could only be like the one that Descartes had had in mind, viz. an immaterial substance. Descartes' conception was, however, absurd because there were, e.g., no criteria for the re-identification of immaterial substances, and therefore, "I" could not denote an object. Now, *formally* this argument is perfectly acceptable, i.e., if one does not want to accept the conclusion as true, one must dispute at least one of the argument's premises. Here the first premise obviously presents itself because there are possibly better candidates for the reference of "I" than immaterial substances...

---

[52] "When, outside philosophy, I talk about myself, I am simply talking about the human being, Anthony Kenny, and my self is nothing other than myself. It is a philosophical muddle to allow the space which differentiates 'my self' from 'myself' to generate the illusion of a mysterious metaphysical entity distinct from, but obscurely linked to, the human being who is talking to you." (Kenny, *The Self*, p. 4)

[53] Anscombe, The First Person, p. 32.

Anscombe's teacher, Wittgenstein, supported a more moderate view for some time. He believed that "I" only sometimes refers, but not always. Wittgenstein thought that one could and should distinguish between two uses of "I," namely, on the one hand, its use as object and on the other hand its use as subject. He more or less contented himself, however, with pointing out the difference between these two uses with the help of a few examples. "I" is used as object in cases like "I have grown ten centimeters" or "I am wounded". What it means to use "I" as subject is to be made clear by examples like "I am in pain" and "I believe that it will rain".

Wittgenstein describes the difference between the two uses of "I" thus: when "I" is used as object, the recognition of a particular person is involved, which is not the case when "I" is used as subject. He who says "I am wounded" must first of all identify an object, namely the object of which he wants to say is wounded, as the object which he wants to refer to by the word "I".

That a recognition is necessary with the use of "I" as object, has as a concequence that such judgements are liable to a certain kind of error. One could be mistaken and falsely believe to be identical to the person one knows to be wounded. Imagine, for example, that you are involved in a car accident with other people. You wake from unconcsiousness, are in terrible pain, realize that some of your limbs are tangled up with those of the others in the accident, and you see that someone's leg is badly wounded. Now you could assume that it is *you* who is wounded – and in this you *could* be wrong, because it could be that your arm is indeed broken, but that your legs have remained unharmed.

Such an error based on a misidentification relative to the respective subject is, according to Wittgenstein, only possible when "I" is used as object. If "I" is used as subject this kind of a mistake cannot happen. It makes sense to ask "Is it really *me* who has grown ten centimeters?" but not "Is it really *me* who is in pain now, or is it my neighbor?"[54]

Wittgenstein's thesis, therefore, states: "I" refers to an object when it is necessary for the correct use of "I" to identify an object, but that is not always the case.

I shall just offer one argument which contradicts not only the radical thesis that "I" *never* refers, but the moderate thesis as well which states that "I" *sometimes* refers: if the sentence "*I* am in pain", uttered by Jack, is true, then the sentence "*Jack* is in pain" is true, just as the sentence "*He* is in pain" is also true when uttered by a third person with respect to Jack. It is difficult not to come to the conclusion that the fact that these sentences

---

[54] Cf. "[...] Mrs Gradgrind's bizarre response when asked on her sickbed whether she was in pain: 'I think there's a pain somewhere in the room, but I couldn't positively say that I have got it'." (Lowe, Self, Reference and Self-reference, p. 21; there Lowe quotes Charles Dickens, *Hard Times*, p. 224).

all have the same truth-value results at least partially from "I", "Jack", and "he" all referring to the same object.[55]

The expression "I" is, therefore, a referring expression; it seems that there is indeed something, *of which* one has a consciousness when one has a consciousness of oneself. But *what* does "I" refer to? In trying to answer this question, one must be careful not to follow the old routes anew which I have just described. It would be easy to say: "I" refers to me, to my self – and there we would be again right back where we started from...

Perhaps it is not so easy to answer the question "What does 'I' refer to?" abstractly. Let us, therefore, take a look at three examples: in the sentence "I have grown 10 centimeters" it seems easy enough – "I" refers 'obviously' to my body since it is indeed my body which has grown. And what about "I am writing an article about self-consciousness"? It is definitely not only my body which is engaged in this activity... As a third example "I believe that two times two is four" should do. Is it my body which believes that? Is my body even involved ? Or is not the mind, the soul, or even the cartesian Ego at issue here?

The examples give the impression that one must differentiate between three possible referents of "I",[56] namely the body, the mind and a *mixtum compositum* of both,[57] and the object to which "I" refers depends on the respective predicate which is attributed to the object.

If, however, you believe that *I* believe that two times two is four, if you believe that *I* have written an article, and if you doubt that *I* have (recently) grown 10 centimeters, do you not each time refer to the *same* object? Although I did indeed suggest that the answer to this question could only be "yes", it remains unanswered as to what kind of an object it is. One proposed solution – which I find plausible – comes from the Oxford philosopher Peter Strawson: In his opinion "I" always refers to a *person*.[58] For Strawson the term "person" is a primitive term and is (therefore) not further analysable:[59]

> "The concept of a person is logically prior to that of an individual consciousness. The concept of a person is not to be analysed as that of an animated body or of an embodied anima. [...] A person is not an embodied ego, but an ego might be a disembodied person, retaining the logical benefit of individuality from having been a person."[60]

So it is persons who use the word "I" and to whom "I" refers. One can attribute physical as well as psychological characteristics to them, and there are

[55] McGinn, *The Character of Mind*, p. 103.

[56] Cf. Wittgenstein's above-mentioned view concerning "I" as object! Cf. also Wittgenstein, *Philosophical Remarks*, §55.

[57] Cf. Strawson, Reply to Mackie and Hidé Ishiguro, p. 268, and Mackie, The Transcendental 'I', pp. 48–61.

[58] Cf. Strawson, *Individuals*, p. 103.

[59] Cf. Strawson, *Individuals*, pp. 101 f.

[60] Cf. Strawson, *Individuals*, p. 103.

(sufficient) criteria to identify and recognize them. As you probably recall, the term "person" has already come up once. It was Kant who maintained that the person is the subject of empirical self-consciousness, i.e. the consciousness which one has of one's own experiences.

Now, is every kind of knowledge which one has in reference to the person who one is a component or expression of self-consciousness? No. I can, for example, know that the next lottery-winner will also receive a Mercedes, and it is possible that I shall be the next lottery-winner. Then I have knowledge about the person who I am, but I am lacking self-consciousness simply because I do not know that *I* am the one who will win the lottery. One could present another reason for denying me self-consciousness in this case: To know that someone is going to receive a Mercedes is not knowledge concerning one of one's own present psychological states; yet was self-consciousness not to be just this kind of knowledge?

Even if we accept this restriction, it won't help us further. Imagine the following case: Anna lies in hospital after an accident; she is suffering from total amnesia, is in terrible pain, and has just woken out of a coma. She hears a nurse tell a doctor: "Anna is in terrible pain". Therefore, the patient believes or knows that Anna is in terrible pain. She also has knowledge of a psychological state which is even her own psychological state, but still this is not an instance of self-consciousness because due to her amnesia Anna does not know what she would report in direct speech by "I am this Anna".[61]

What Anna is lacking in this example is something indispensable for every episode of self-consciousness. If I display self-consciousness, it is not enough that I know something about the person who, as a matter of fact or as it just happens, I am; I must know it about the person who I am *in a certain way*; I must know that *I, myself*, am the person in question.[62]

We are still in search of the nature of self-consciousness. As we have just seen, someone who simply has knowledge about the person he is, does not have self-consciousness in *this* respect, whereas someone else who has knowledge about *himself* in this respect possesses self-consciousness. What is the difference between these two persons?

The American philosopher John Perry presents the following example:

"I once followed a trail of sugar on a supermarket floor, pushing my cart down the aisle on one side of a tall counter and back the aisle on

---

[61] I don't want to doubt, however, that Anna knows that she is in pain, since she is feeling the pain. *That* is a case of self-consciousness, but not the knowledge she acquired in the way described and which she would express by "Anna is in pain.".

[62] This expression, "I myself", has been called a 'quasi-indicator'. As to how exactly this quasi-indicator is to be understood and what its distinctive features are, in comparison to other indicators, has been investigated in detail by the American philosopher Hector-Neri Castañeda. (Cf. especially his seminal articals "'He': A Study in the Logic of Self-consciousness" and "Indicators and Quasi-Indicators".)

the other, seeking the shopper with the torn sack to tell him he was making a mess. With each trip around the counter, the trail became thicker. But I seemed unable to catch up. Finally it dawned on me. I was the shopper I was trying to catch."[63]

During his first rounds around the counter Perry knows something about the person he is, viz. he knows of one of the customers that he is losing sugar. It is only later that he realizes that it is he himself who is losing sugar, and immediately his behavior changes. First he tries to catch up with the other customer, but as soon as he realizes that he himself is the cause, he stops his search and instead takes care of the damaged bag of sugar in his own shopping cart. Here self-consciousness displays itself in behavior.[64]

Therein exists a difference between someone who in a certain respect does not possess self-consciousness in the sense of self-knowledge (this is Perry at the beginning of the adventure) and someone who does possess self-consciousness (that is Perry at the end of his experience). Not everyone, however, who – and not everything, which – behaves like Perry in this example has *therefore* self-consciousness. Self-consciousness is *expressed* through behavior, but is more than just a way of behaving. How is it more? As already indicated, self-consciousness definitely calls for an epistemological component: one must refer to oneself knowing that it is *oneself* one is referring to. However, does knowing this really help? It would be wise to let Kant have the last – admittedly difficult – word on this topic:

> "How it is possible that I, I who think, can be an object (of intuition) for myself and so distinguish me from myself, is absolutely impossible to explain, although it is an indisputable fact; but it indicates a faculty so much superior to all sensible intuitions that, since that faculty is the possibility of an understanding, it has as its consequence the complete separation from beasts, to which we have no reason to attribute the ability to say I to itself [...] "[65, 66]

---

[63] Perry, The Problem of the Essential Indexical, p. 3.

[64] Cf. Anscombe, The First Person, p. 64.

[65] "Wie es möglich ist, daß ich, der ich denke, mir selber ein Gegenstand (der Anschauung) sein, und so mich von mir selbst unterscheiden könne, ist schlechterdings unmöglich zu erklären, obwohl es ein unbezweifeltes Faktum ist; es zeigt aber ein über alle Sinnenanschauungen so weit erhabenes Vermögen an, daß es, als der Grund der Möglichkeit eines Verstandes, die gänzliche Absonderung von allem Vieh, dem wir das Vermögen, zu sich selbst Ich zu sagen, nicht Ursache haben beizulegen, zur Folge hat [...] " (Kant, Welches sind die wirklichen Fortschritte, die die Metaphysik seit Leibnizens und Wolffs Zeiten in Deutschland gemacht hat?, A 35 f; my translation into English)

[66] For helpful and critical comments on an earlier version of this article I would like to thank S. Günther, P. Krüger and J. Kulenkampff. Many thanks are also due to A. Blizzard, who did most of the translating into English.

References

Anscombe, E.G.M. (1975): The First Person, in: S. Guttenplan (ed.), *Mind & Language* (Oxford University Press, Oxford) pp. 45–65

Aristotle: *The Works of Aristotle*, translated into English under the Editorship of W. D. Ross, 12 Vols., (Vol. 9: Ethica Nicomachea by W. D. Ross, Magna Moralia by St. G. Stock, Ethica Eudemia & De Virtutibus et Vitiis by J. Solomon) (Oxford University Press, London) 1908–1954

Armstrong, D..M. (1968): *A Materialist Theory of Mind* (Routledge & Kegan Paul, London and Henley)

Baldwin, J.M. (ed.) (1925): *Dictionary of Philosophy and Psychology*, 3 Vols. (Peter Smith, Gloucester, Mass.) corrected edn. 1925, reprinted 1960

Carriero, J.P. (1986): The Second Meditation and the Essence of Mind, in: A. Oksenberg Rorty (ed.), *Essays on Descartes' Meditations* (University of California Press, Berkeley and Los Angeles) pp. 199–221

Castañeda, H.-N. (1966): 'He': A Study in the Logic of Self-consciousness, *Ratio 8*, pp. 130–157

Castañeda, H.-N.(1967): Indicators and Quasi-Indicators, *American Philosophical Quarterly 4*, pp. 85–100

Descartes, R. (1931a): *The Philosophical Works of Descartes*, ed. and transl. by E. Haldane and G. R. T. Ross, 2 Vols. (Cambridge University Press, Cambridge) reprinted with corrections 1976

Descartes, R. (1931b): Meditations on First Philosophy, in: R. Descartes, *The Philosophical Works of Descartes*, ed. and transl. by E. Haldane and G. R. T. Ross, Vol. 1 (Cambridge University Press, Cambridge) reprinted with corrections 1976, pp. 131–199

Descartes, R. (1931c): Discourse on the Method of Rightly Conducting the Reason, in: R. Descartes, *The Philosophical Works of Descartes*, ed. and transl. by E. Haldane and G. R. T. Ross, Vol. 1 (Cambridge University Press, Cambridge) reprinted with corrections 1976, pp. 79–130

Descartes, R. (1931d): The Principles of Philosophy, in: R. Descartes, *The Philosophical Works of Descartes*, ed. and transl. by E. Haldane and G. R. T. Ross, Vol. 1 (Cambridge University Press, Cambridge) reprinted with corrections 1976, pp. 201–302

Dickens, Ch. (1969): *Hard Times* (Penguin, Harmondsworth)

Duden. Das große Wörterbuch der deutschen Sprache, (1980): 8 Bd., hrsg. und bearb. vom Wissenschaftlichen Rat und den Mitarbeitern der Dudenredaktion unter der Leitung von G. Drosdowski (Bibliographisches Institut, Mannheim)

Fichte, J. G. (1937): *Wissenschaftslehre 1798 nova methodo*, in: J. G. Fichte, *Schriften aus den Jahren 1790–1800*, hrsg. von H. Jacob (Juncker und Dünnhaupt, Berlin) (= J. G. Fichte, *Nachgelassene Schriften*, Vol. 2 (Juncker und Dünnhaupt, Berlin) 1937)

Frege, G. (1977): Thoughts, in: G. Frege, *Logical Investigations*, ed. with a preface by P. T. Geach, transl. by P. T. Geach and R. H. Stoothoff (Blackwell, Oxford) pp. 1–30

Glock, H. (1995): Innen und Außen: "Eine ganze Wolke Philosophie kondensiert zu einem Tröpfchen Sprachlehre", in: E. von Savigny & O. R. Scholz (Hrg.), *Wittgenstein über die Seele* (Suhrkamp, Frankfurt/M.)

60   Thomas Spitzley

Hume, D. (1987): *A Treatise of Human Nature*, ed. by L. A. Selby-Bigge, second edition, with text revised and notes by P. H. Nidditch (Clarendon Press, Oxford)

Kant, I. (1933): *Critique of Pure Reason*, transl. by Norman Kemp Smith (Macmillan, London and Basingstoke) second impression with corrections 1973

Kant, I. (1983): *Welches sind die wirklichen Fortschritte, die die Metaphysik seit Leibnizens und Wolffs Zeiten in Deutschland gemacht hat?*, in: I. Kant, *Werke in zehn Bänden*, Bd. 5, hrsg. von W. Weischedel (Wissenschaftliche Buchgesellschaft, Darmstadt)

Kenny, A.J.P. (1988): *The Self* (The Aquinas Lecture, 1988), (Marquette University Press, Milwaukee)

Lowe, E.J. (1993): Self, Reference and Self-reference, *Philosophy 68*, pp. 15–33

Mackie, J. L. (1980): The Transcendental 'I', in: Z. van Straaten (ed.), *Philosophical Subjects. Essays presented to P. F. Strawson* (Clarendon Press, Oxford) pp. 48–61

McGinn, C. (1982): *The Character of Mind* (Oxford University Press, Oxford)

The Oxford English Dictionary (1989): prepared by J. A. Simpson and E. S. C. Weiner, 20 Vols., second edn. (Clarendon Press, Oxford)

Perry, J. (1979): The Problem of the Essential Indexical, *Noûs 13*, pp. 3–21

Ritter, J. (Hrg.) (1971): *Historisches Wörterbuch der Philosophie*, Bd. 1 (Wissenschaftliche Buchgesellschaft, Darmstadt)

Ryle, G. (1970): *The Concept of Mind* (Penguin, Harmondsworth)

Scholz, H. (1931/32): Augustin und Descartes, *Blätter für deutsche Philosophie 5*, pp. 405–423

Shoemaker, S. (1986): Introspection and the Self, *Midwest Studies in Philosophy X*, pp. 101–120

Strawson, P. F. (1977): *Individuals. An Essay in Descriptive Metaphysics* (Methuen, London)

Strawson, P. F. (1978): *The Bounds of Sense* (Methuen, London)

Strawson, P. F. (1980): Reply to Mackie and Hidé Ishiguro, in: Z. van Straaten (ed.), *Philosophical Subjects. Essays presented to P. F. Strawson* (Clarendon Press, Oxford) pp. 266–273

Thomas Aquinas (1952–54): *Truth (de Veritate)*, trans. R. W. Mulligan, J. V. McGlynn & R. W. Schmidt (Henry Regnery Co., Chicago)

Tugendhat, E. (1979): *Selbstbewußtsein und Selbstbestimmung* (Suhrkamp, Frankfurt/M.)

Wilson, M. D. (1991): *Descartes* (Routledge, London and New York)

Wittgenstein, L. (1958a): *Philosophical Investigations*, transl. by G. E. M. Anscombe (Blackwell, Oxford) second edn.

Wittgenstein, L. (1958b): The Blue Book, in: L. Wittgenstein, *The Blue and Brown Books* (Blackwell, Oxford)

Wittgenstein, L. (1968): Notes for Lectures on 'Private Experience' and 'Sense-Data', ed. by R. Rhees, *The Philosophical Review 77*, pp. 275–320

Wittgenstein, L. (1975): *Philosophical Remarks*, ed. by R. Rhees and transl. by R. Hargreaves and R. White (Blackwell, Oxford)

Meaning Skepticism and Cognitive Science*

Felix Mühlhölzer

Philosophisches Seminar, Georg-August-Universität Göttingen,
Humboldtallee 19, D-37073 Göttingen, Germany

In Chap. 4 of his book *Knowledge of Language*, Noam Chomsky deals with the
meaning skepticism developed in Saul Kripke's much discussed book *Wittgenstein on Rules and Private Language*.[1] Chomsky admits that "[o]f the various
general critiques that have been presented over the years concerning the program and conceptual framework of generative grammar, this [i.e. Kripke's]
seems to me the most interesting." (Chomsky 1986, p. 223) Nevertheless, he
believes himself to be capable of refuting Kripke's analysis simply by referring to the possibility of constructing an empirical theory about our cognitive
capacities, in particular about our semantic capacities, and especially about
our semantic competence with respect to the word "plus" (or the sign "+"),
which serves as the main example in Kripke's book.

In this contribution I take Chomsky as a representative of cognitive science, and by discussing his reply to Kripke, I intend to explore the possible
merits and demerits of cognitive science in general, as it attempts to deal
with Kripke's meaning skepticism.[2] If I am not mistaken, Chomsky is the
only cognitive scientist who up to now has really tried to seriously answer
the Kripkean skeptic; at the same time Chomsky's answer has met with
only little response in the philosophical literature.[3] I find this astonishing

---

* This article grew out of the paper "Über eine mögliche Inkonsistenz in Chomskys
Auffassung von Sprachregeln", presented at the congress *Analyomen–Perspectives
in Analytical Philosophy* in October 1991 at the University of Saarbrücken (published as Mühlhölzer/Emödy 1994) which I wrote jointly with Marianne Emödy.
Marianne Emödy and I have had innumerable discussions on the present topics
over the past eight years, and I am most thankful to her for them. Furthermore
I am grateful to Noam Chomsky, Christoph Demmerling, Andreas Kemmerling,
Hilary Putnam, Heinz-Jürgen Schmidt and Arnim von Stechow for valuable comments on previous drafts of this article and to Mark Helme and Kimberly Miller
for correcting and improving my English.

[1] In what follows I will presuppose familiarity with Kripke's book.

[2] When I refer to something as 'Kripke's meaning skepticism', 'Kripke's skeptical
argument', and so on, I merely mean to label the idea in question as one that
Kripke presents in his interpretation of Wittgenstein, without suggesting that
he himself fully endorses it. Furthermore, since the subject of the present paper
is Chomsky's view – and the view of cognitive scientists in general – and not
Wittgenstein's, I will remain silent about questions concerning the adequacy of
Kripke's interpretation of Wittgenstein.

[3] Ignoring reviews of Chomsky 1986, the only exceptions known to me are Wright
1989 and Bilgrami 1992. Wright's discussion of Chomsky's position, however,

and not at all justified, and in what follows I want to thoroughly scrutinize the Chomskyan point of view from a philosophical perspective. It is an important characteristic of Chomsky's viewpoint that he considers linguistics, and cognitive science in general, simply as belonging to the natural sciences, ultimately as a part of biology, say; and in his linguistic work he always stresses the similarities between linguistics and the natural sciences because he thereby expects an advancement of linguistics. This is without doubt a scientifically fruitful procedure. In contrast to Chomsky, however, I will focus, at the end of this contribution, on the dissimilarities between linguistics and the natural sciences because I am interested in the limitations of the latter. This more philosophical orientation leads to questions which, in my opinion, have not really been done justice in Chomsky's writings and in the writings of cognitive scientists in general. I do not believe that a naturalist like Chomsky can tell us the *whole* story about language. On purely scientific grounds, Chomsky himself doesn't believe that either.[4] An important task, then, is to specify the real gaps and the real problems of his position, and I think that not only scientific but, despite Chomsky's defense, also philosophical considerations like the Kripkean ones are helpful in opening our eyes to some of these gaps and problems.

In the first section, after sketching Kripke's skeptical problem about meaning, I will expound Chomsky's reply to Kripke and commit myself to a specific interpretation of it which will be the basis of the subsequent investigation. In Sect. 2, I will discuss what Kripke calls "normativity", which plays a decisive role in his book. It will turn out that normativity considerations alone are not sufficient to undermine Chomsky's position. In the final section, then, I will argue that a definite problem in Chomsky's position can be discovered quite

is very meagre. Within barely half a page (p. 236), Wright criticizes Chomsky on two counts: that he doesn't do justice to (a) the 'normativity' inherent in meaning and (b) our self-knowledge with respect to what we mean; but apart from that, Wright simply pursues his own thoughts concerning the question of the factuality of meaning. In what follows, I will argue, in Section 2, that Chomsky has a sufficient defense against the normativity-objection, and, in Section 3, that Chomsky in fact may stumble over considerations about self-knowledge, but only if these considerations are backed up by certain reflections concerning the status of Chomsky's linguistics. In Bilgrami's rich book, too, Chomsky's answer to Kripke is dealt with only in passing and, strictly speaking, only implicitly. In contrast to Wright, Bilgrami rejects the Kripkean normativity-objection and (if I understand him correctly) he thinks that meaning-skeptical considerations are not likely to undermine Chomsky's position anyway. I agree with Bilgrami on the first point (Sect. 2) but not on the second (Sect. 3).

4 See, e.g., Chomsky 1995, p. 27: "[...] general issues of intentionality, including those of language use, cannot reasonably be assumed to fall within naturalistic inquiry." Chomsky reaches conclusions like this by estimating, as a scientist, the possibilities of a naturalistic approach. In what follows, I will try to present, instead, some *philosophical* arguments leading to the same result.

directly in his assumption that a semantic theory should be an empirical one, comparable to theories in physics, chemistry, or biology.

1   Chomsky's Reply to Kripke's Skeptical Problem

In his book on Wittgenstein Kripke invents a skeptic who, from the metalevel of our present speech – the meaningfulness of which is not questioned – asks about the meaning of our previous utterances. What, for example, did we mean up to now by the word "plus" or by the plus sign? Kripke's skeptic claims that there is no fact – no 'meaning-fact', as it were – according to which by "plus" we meant the plus function as opposed to a host of other possible functions. Let us call these other functions that the skeptic can put forward here "quus functions".[5] In his book Kripke is searching for meaning-facts that could refute the skeptic, but despite considerable, sophisticated efforts he doesn't find any. What he finds is an enormous, in fact an infinite set of possible quus functions, that is, an enormous indeterminacy in what we could have meant by "plus". The skeptic seems to be right in the end: There is no fact of the matter as to what we meant by "plus". The problem at stake here is a problem of determinacy (or of uniqueness): Faced with the skeptic's challenge, how can we save the determinacy (or uniqueness) of what we meant by our terms? That is, how can we save the determinacy of their extensions? Let us ignore for the moment the fact that Kripke tries to mould this problem into the form of a paradox.

Determinacy problems like the one just outlined are nothing new. For example, the philosophy of W.V. Quine, notoriously, is permeated with them, in such a way in fact that it could readily be called a "philosophy of indeterminacy". Kripke's and some of Quine's indeterminacies seem to arise if we confront intentional categories – like "to mean" in Kripke or "to refer to" in Quine[6] – with the domain of facts (which in Quine's case is identified with the domain of scientific facts). It seems as if the domain of facts is not able to confer determinacy upon the domain of the intentional, at least not if we want to understand the intentional categories in a substantive way.

The question of whether the domain of facts is able to confer determinacy upon the domain of the intentional has substance only if one refrains from postulating 'intentional facts' in their own right, like, for example, an autonomous 'fact of meaning plus by "plus"'. So, as already mentioned, Quine only accepts facts stated by the natural sciences, ultimately even only facts stated by physics; and Kripke, who does not explicate his notion of a fact, should be interpreted in a way which in any case leaves the said question its

---

[5] I deviate slightly from Kripke's use of the expression "quus function". Kripke uses it as a singular term referring to a specific function, while I will use it as a general term applying to all the functions that serve the skeptic's purpose.

[6] Quine expounds his view about reference in, among others, Quine 1961, Quine 1981, Essay 1, and Quine 1992, Chap. II.

substance. Kripke is vague in this respect, but his vagueness is a problem-oriented and therefore, it seems to me, a legitimate one. He considers, e.g., facts concerning our dispositions to verbal behavior, facts revealed by intro-spection, neurophysiological facts, and facts concerning our functional orga-nization. One may extend this list, but the crucial point is that all these facts must be characterized in a way that does not beg the question. Thus, one is not allowed to characterize any of them by phrases like "the fact so-and-so constituting our meaning plus by 'plus'". Otherwise, one would leave the skeptic's challenge unanswered – and I find it obvious that this challenge should be answered, either by a 'straight' or by a 'skeptical' solution.[7]

Speaking very roughly, we can now say that the indeterminacy which comes to light when the intentional is measured against the factual leads Quine to sacrifice the intentional in favor of the factual. In contrast to Quine, Kripke, when giving his skeptical solution to the skeptical problem, accepts the said indeterminacy only initially, but then, as it were, sacrifices the facts in favor of the intentional, especially in favor of meaning. According to the skeptical solution, meaning should not be seen as constituted by 'meaning-facts' but as finding its expression in certain structures of our language games.

What does Chomsky say to all that? In his reply to Kripke, he brings into play not the structures of our language games but the structure of our brain. Surely, our linguistic utterances and our reactions to linguistic utterances are to a considerable extent determined by the structure of our brain (so that the structures of our language games are to a considerable extent determined by the structure of our brain, too). Seen in this way, it does not in fact seem implausible to study the structure of the brain when one is interested in linguistic meaning and its determinacy.

Now, if I understand Chomsky correctly, he wants to give a straight so-lution to the skeptical problem, i.e., he really wants to present a specific fact constituting plus- or quus-meaning which will guarantee the determinacy of the extension of the word "plus". His solution is roughly as follows: Let us regard a human being, say Jones, or at least his brain, as a sort of computer and find out which program is implemented in this computer. That Jones means plus – or quus – by "plus" should be represented by a specific compo-nent of this program. This component is causally relevant to Jones's use of the word "plus", in quite a similar way as an ordinary computer program is causally relevant to the computer's input–output behavior. That Jones means the plus function – or a certain quus function – by "plus", then, consists in

---

[7] These two notions are introduced by Kripke as follows: "Call a proposed solution to a sceptical problem a *straight* solution if it shows that on closer examination the scepticism proves to be unwarranted [...]. A *sceptical* solution of a sceptical philosophical problem begins on the contrary by conceding that the sceptic's negative assertions are unanswerable. Nevertheless our ordinary practice or belief is justified because – contrary appearances notwithstanding – it need not require the justification the sceptic has shown to be untenable." (Kripke 1982, p. 66)

nothing else but the fact that the said component of the program with the said causal relevance is to be found in Jones's brain. So it will be the scientific theory describing all that – a theory that surely must be a mathematical one in part – which will inform us about the facts that distinguish meaning the plus function (component of the program representing the plus function) from meaning a certain quus function (component of the program representing this quus function).

Thus, Chomsky, in a manner well-tried in the natural sciences, simply devises a model.[8] It is a model of our (or Jones's) cognitive and, in particular,

----

[8] I have drawn this model mainly from pp. 236–239 and 244 of Chomsky 1986. On p. 244 Chomsky states his methodological stand with respect to constructing such a model: "My assumption so far (as in earlier work) has been that we are entitled to propose that the rule R [the rule of addition associated with "plus", say] is a constitutive element of Jones's language (I-language) if the best theory we can construct dealing with all relevant evidence assigns R as a constituent element of the language abstracted from Jones's attained state of knowledge. Furthermore, we are entitled to propose that Jones follows R in doing such-and-such [...] if, in this best theory, our account of his doing such-and-such invokes R as a constituent of his language."

Note that Chomsky here uses the word "language" in a very technical sense, namely, in the sense of what he calls an *I-language* (which is short-hand for "internalized language"). In Chomsky's own words, the I-language "is some element of the mind of the person who knows the language, acquired by the learner, and used by the speaker-hearer" (Chomsky 1986, p. 22; of course, by "mind" Chomsky here means the program, i.e., the software, of our computer-brain). The I-language must be distinguished from the so-called *E-language* (which is shorthand for "externalized language"). The E-language is nothing mental, as the I-language is, but "a collection of actions, or utterances, or linguistic forms (words, sentences) paired with meanings" which is to be "understood independently of the properties of the mind/brain." (Chomsky 1986, p. 19f.) Chomsky thinks that the notion of an E-language is unfit for scientific purposes and that the linguistically fruitful sense of "language" is that of an I-language. If Chomsky uses the term "language", it is mostly in the sense of an I-language.

The notion of an I-language is part of Chomsky's *naturalism*. A naturalist is somebody who accepts as answers to all theoretical questions, even to all theoretical philosophical questions, only those given by the natural sciences and who tends to ignore any question that defies scientific treatment; where, according to Chomsky, the natural sciences should be characterized as "simply follow[ing] the post Newtonian course, recognizing that we can do no more than seek the best theoretical account of the phenomena of experience and experiment, wherever the quest leads." (Chomsky 1995, p. 39) To a naturalistically minded linguist like Chomsky it is a very attractive idea to consider the human brain as a sort of computer – or, more precisely, as the hardware of a computer – and the human mind as the software of this computer. Of course, this computer, including hard- *and* software, has not been made by man but by nature itself. According to this point of view, it seems that one can be a naturalist and at the same time accept the human mind as something relatively independent of the details of the human

semantic capacities. And Chomsky seems to believe that this model can serve as a straight solution to the skeptical problem: The fact of meaning plus, for example, would essentially be constituted by a component of our computer-brain's program that represents the plus function.

Notice that there is also another way of looking at Chomsky's model. One is not necessarily forced to understand it as presenting a straight solution. Generalizing the Kripkean notion of a skeptical solution by giving it a purely negative formulation, from now on when I speak of a *skeptical solution*, I will mean any solution or dissolution of the skeptical problem that does not, whether explicitly or implicitly, make use of a substantive conception of 'meaning-facts which are capable of uniquely determining the extensions of our terms'.[9] Then, Chomsky's model can be interpreted as a skeptical solution in this sense as well, if one considers it not as a model of extension determination but merely as a model of certain mechanisms that are causally relevant to our linguistic behavior,[10] whereby the issue of how

---

brain. Mental states and properties may be considered as determined by certain functional roles within the neurophysiological happenings of the brain, and the material details of these happenings may be deemed to be irrelevant as long as the functional roles remain the same. An I-language, then, is nothing but a specific component of the software of our computer brain, which Chomsky sometimes calls the "language faculty" or even the "language organ".

[9] Kripke tends to give a more specific, positive characterization of a skeptical solution, involving assertability conditions of sentences in contrast to the truth conditions which are supposed to be stated by a straight solution (see Kripke 1982, pp. 73–78). In what follows, I will say nothing about this more specific notion and I will always stick to my more general notion of a skeptical solution since it appears to me more useful. Notice, by the way, that other authors have also been driven to develop their own notion of a skeptical solution according to their own purposes. Bilgrami, e.g., calls "skeptical solution" any solution to the skeptical problem that "proposes [...] not to look merely at the agent for the solution, but to look at the community in which he lives and speaks." (Bilgrami 1992, p. 87) This notion serves Bilgrami's purposes very well, which, however, differ considerably from mine. (Incidentally, the solution which Bilgrami himself gives to Kripke's skeptical problem – and which I cannot go into here; see Bilgrami 1992, pp. 92–99 – is a skeptical solution in my sense, as can be seen, e.g., from the following statement on p. 97: "For me, it is quite true that there is no fact of the matter about whether any of us right now is following the ordinary rule for our terms or whether for each of our terms we are following the sort of bizarre rules [entertained by a Kripkean skeptic].")

[10] Chomsky sometimes seems to deny the causal relevance of rules (like the rule of addition), for example on p. 260 of Chomsky 1986, where he writes: "Our behavior is not 'caused' by our knowledge, or by the rules and principles that constitute it. In fact, we do not know how our behavior is caused, or whether it is proper to think of it as caused at all." But this means overshooting the mark, and it contradicts what Chomsky writes on p. 244, namely, that "if [the rule] R is a constituent element of the initial state as determined by our best theory, and invoking R is part of our best account of why the attained state has such and

the extension determinations may be accomplished will be trivialized by a simple 'disquotational' or 'deflationary' view of reference.

Let me explain this solution in a little more detail. In the case of the word "plus", for example, Chomsky could simply be content with referring to all the scientifically specifiable mechanisms that are causally relevant to our use of this word. He could call *them* "the meaning of 'plus'" – where "meaning" is used here in the sense of "intension"[11] – and he could settle, then, the issue of extension-determination simply and exclusively by the disquotational schema

(D) "plus" means (i.e., refers to) plus.

This move appears to dispose of all the Kripkean skeptical worries at one blow. For, the mechanisms which are causally relevant to my *present* use of the term "plus" are certainly the same as the mechanisms causally relevant to my *past* use of this term. Therefore, without being guilty of a fallacy of equivocation with respect to "plus", I am allowed to consider the schema (D) as settling what I mean by my present as well as what I meant by my past use of "plus". Consequently, what I meant by "plus" in the past is the *same* as what I mean now. But what I mean now is *presupposed* to be determined,[12] and so what I meant in the past must be determined, too – and that in the trivial sense of (D). It appears, then, that all the objections put forth by Kripke's skeptic will become idle.

I don't think, however, that this is the position which Chomsky himself had in mind when writing his reply to Kripke. Therefore, in what follows, I

---

such properties that then enter into behavior, we are entitled to propose that R has 'causal efficacy' in producing these consequences." The reason for the apparent contradiction may be that, occasionally, Chomsky may have reservations about speaking of mechanisms. In fact, if with causality we associate mechanisms and if by mechanisms we understand only physical (or, in the Chomskyan context, physiological) mechanisms, then Chomsky is *not* dealing with mechanisms and ipso facto not with causality. However, we may well permit ourselves to use the concept of a mechanism also in a more abstract sense, to include mechanisms that are described purely functionally, and then Chomsky is dealing with mechanisms in this sense, and of course these mechanisms in general *are* causally relevant to our behavior. (Another danger of the notion of a 'mechanism' lies in its deterministic undertone. Nothing would be further from Chomsky's view than a deterministic picture of human faculties and performance; see, e.g., Chomsky 1986, pp. 222f.)

[11] Alternatively, Chomsky could, like Quine in his article "Use and Its Place in Meaning", simply say: "Well, we can take the [mechanisms] and let the meaning go." (Quine 1981, p. 46) I do not here take a stand on which of these two routes should be chosen, since this seems to me irrelevant in the present context.

[12] Without this presupposition Kripke's skeptical problem couldn't even be formulated; see Kripke 1982, pp. 11f. Quine expresses presuppositions of this sort in the following, emotionally appealing way: "[We] acquiesc[e] in our mother tongue and tak[e] its words at face value." (Quine 1969, p. 49) Of course, in the Kripkean context 'our mother tongue' has to be equated with 'our present tongue'.

will consider Chomsky's model of rule following exclusively as aiming at a
straight solution of the skeptical problem, i.e., at a solution which presents a
specific fact constituting plus- or quus-meaning such that this fact – and not
merely a disquotational schema – guarantees the determinacy of the extension
of the word "plus". The alternative interpretation as a skeptical solution needs
a separate investigation which I will leave for another occasion.[13] From now
on, when I refer to something as 'Chomsky's model' or 'Chomsky's solution',
etc., this is always to be understood in the sense of a straight solution.

2   Normativity

The question now is whether Chomsky's model really adequately represents
our capacity to mean something by a word. In the literature we can find
all kinds of possible objections to Chomsky's view; for example, that it is
inconsistent with meaning holism,[14] i.e., with the fact (if it is one) that there
are no 'meanings' in the sense of isolable items assigned to single words
or single sentences, but rather that meaning is something pertaining only
to a whole, interconnected system of words and sentences. This objection
has much plausibility, I think, if it is restricted to our non-logical and non-
mathematical vocabulary, but it loses its plausibility in the case of a word
like "plus", which I am considering here. (Maybe, the natural numbers are
the work of God, and our meaning plus by "plus" is too.) So in what follows
I will simply ignore the issue of meaning holism.[15]

In this section I will focus on a further possible objection to Chomsky,
one which plays an important role in Kripke's book. It concerns what Kripke
calls "normativity".[16] I want to show that this objection, taken in isolation

---

[13] A skeptical solution along the lines sketched above has already been suggested in
Mühlhölzer/Emödy 1994, pp. 481f. More recently, and in a more elaborate form,
a solution of this sort has been argued for in Horwich 1995. The similarity of it to
what one knows from Quine may evoke the suspicion, however, that it will lead
us in the end also to the Quinean indeterminacy ('inscrutability') of reference.
(In Mühlhölzer 1995 I have put together reflections in the light of which this
Quinean thesis may appear almost inevitable.) As a reply to Kripke's skeptic
this may look a bit strange.

[14] See, e.g., Putnam 1988, pp. 8–11.

[15] The specific case of "plus" also does not seem open to objections that contemplate
twin earths and the like (see Putnam 1988, pp. 30–33), since in this case there do
not seem to be relevant 'contributions of the environment'. – In discussing only
the word "plus", I deliberately restrict myself to a case which is most favorable
to the Chomskyan model. However, we will see that even then serious objections
remain.

[16] Kripke has used this objection mainly against the 'dispositional response' to
meaning skepticism, but it can be raised against Chomsky's response as well;
better still, Chomsky's response can actually be subsumed under the dispositional
response if the latter is formulated in a sufficiently sophisticated way (as has been

from other objections, is not as strong as Kripke deems it to be, and that Chomsky may well have an answer to it.

Following Wittgenstein, Kripke points out that it is certainly essential to our notion of meaning, as well as to notions like "concept" and "rule", that we distinguish between correct and incorrect applications of meaningful words, concepts and rules. Take the word "plus". If we ask, "How much is 68 + 57?", the correct answer is "125", and all other answers are incorrect – provided that by "+" we mean the plus function (and by "68", "57" and "125" the numbers 68, 57 and 125, respectively). And one should think, then, that the corresponding fact as to what we mean by "+" somehow has to 'contain' this normative component, that it somehow brings along a certain standard for 'right' and 'wrong'. Kripke's skeptic, however, cannot find facts of this sort, and he concludes that they simply do not exist. That they are a myth. Kripke expresses it thus: "There is no 'superlative fact' [...] about my mind that constitutes my meaning addition by 'plus' and determines in advance what I should do to accord with this meaning." (Kripke 1982, p. 65) Kripke here alludes to § 192 of Wittgenstein 1953, where Wittgenstein seems to agree with him: "You have no model of this superlative fact."[17]

What, then, about Chomsky's model? It will be acceptable only if it can present facts of just this sort. For example, it must present a fact as to what we mean by "+", which possesses this normative force. Can we discover in Chomsky's model an acceptable standard for correct and incorrect applications of "+"? In a certain sense, we can indeed. For, Chomsky's model allows a clear cut distinction between our *competence*, realized by the (undisturbed) language organ, and our *performance*, that is, our actual linguistic behavior with all its imperfections. The competence–performance distinction in turn allows the distinction between a correct and an incorrect answer to, say, the question "How much is 68 + 57?". The answer is correct if the performance flows from the competence in an undisturbed manner; otherwise, it is incorrect.

---

done, e.g., in Boghossian 1989, pp. 527–540, or in Horwich 1995, pp. 555–557). Many philosophers, Crispin Wright for example (see Wright 1989, p. 236), regard Kripke's normativity objection as convincing.

[17] Wittgenstein is quite specific about this fact. According to him, it should consist in "grasp[ing] the whole use of the word in a flash" (Wittgenstein 1953, § 191). Thus, it has the normative ingredient which we are interested in here, because it somehow contains the whole use – i.e., the whole *correct* use – of the word; but in addition it also involves the *grasping* of this use, i.e., a conscious mental act. In the present Chomskyan context, I shall ignore this additional aspect, since the mental states, etc. which Chomsky deals with are in general hidden from our consciousness. We must allow that the facts we are looking for are not present in our consciousness. (Of course Wittgenstein, notoriously, loathes such hidden things in his philosophy; see Wittgenstein 1953, p. 126: "[...] what is hidden [...] is of no interest to us.")

Notice that there are quite different sorts of possible disturbances. Suppose that Chomsky's model says in fact that by "+" we mean the plus function. Then, first of all, there is the 'disturbance' which consists in our inability to add very large numbers. Though it lies within our *competence* to add any two numbers, however large, the limitations of our organism prevent us from actually *performing* all these additions.[18] This obviously involves a rather ambitious conception of competence, which is, however, necessary if Chomsky really wants to give a straight solution to Kripke's problem. Furthermore, there are of course the mundane disturbances leading to what we call "miscalculations" (including deliberate ones). Finally, our language organ itself may be disturbed, because of brain injury, say, leading to a loss of even our (factual) competence. Even then, however, it seems sensible to distinguish between correct and incorrect answers, since we can measure the answers against what would accord with our undisturbed language organ. In all three cases, therefore, it is the undisturbed language organ, put to use in an undisturbed way, that determines what has to count as 'right' or 'wrong'.

Thus, Chomsky may very well seem to be able to meet Kripke's normativity requirement. Kripke, however, raises the following objection (see Kripke 1982, pp. 27–32): Chomsky's competence–performance distinction is essentially based on the notion of a 'disturbance', but to be able to say what a disturbance is, we must first define what has to count as 'undisturbed'; this, however, in the case of "+", necessarily involves saying what is *meant* by "+"; hence, Chomsky's account is circular. Kripke thinks that we cannot avoid circularities of this sort. He writes: "[...] we might try to specify the [disturbances] to be ignored without presupposing a prior notion of which function is meant. A little experimentation will reveal the futility of such an effort." (Kripke 1982, pp. 30f)

This, however, is too rash. Kripke seems to think that the question: What is an undisturbed state and what isn't?, cannot be an empirical one, but he does not tell us why. In (scientific) reality it does not seem at all sensible to settle questions like these in an *a priori* way. No wonder then that Chomsky

---

[18] However, we are able to construct devices external to our brain – computers, for example – which help us to extend our performance; and there does not seem to be any principled limit to such extensions. Thus, Chomsky thinks it a mistake to consider our brain as a *finite* automaton (at least as a finite automaton "in the sense of the fifties"): "It is the wrong approach because even though we have a finite brain, that brain is really more like the control system for an infinite computer. That is, a finite automaton is limited strictly to its own memory capacity, and we are not. We are like a Turing machine in the sense that although we have a finite control unit for a brain, nevertheless we can use indefinite amounts of memory that are given to us externally to perform more and more complicated computations. [...] We can go on indefinitely, if we just have more and more time, and memory, and so on. We do not have to learn anything new to extend our capacities in this way. This means that we are not finite automata in the sense of the fifties." (Chomsky 1982, p. 14)

simply proceeds in an *empirical* manner as follows: Let us construct a theory about human beings which says, among other things, what 'undisturbed functioning' of their organism is and what it is not. This theory would, first of all, postulate a brain structure that can partly be identified as the realization of a certain program; furthermore, it would show that this program represents a certain function (hopefully, the plus function). Both steps essentially involve mathematics, and especially arithmetic. This theory, then, shows that we are 'programmed' to respond to questions like "How much is 68 + 57?" with (it is to be hoped) the sum and not with the quum. (The quum is the result according to the quus function.) Since the theory is an empirical one, the question as to what 'undisturbed functioning' is becomes an empirical question too: We should view as undisturbed functioning that which our best theory says that undisturbed functioning is. By being an empirical one, a theory of this sort must even be allowed (so it seems) to contain the result that by the word "plus" we do *not* mean plus. Hence, Chomsky's account does not appear circular at all.

One may ask, nevertheless, whether the Chomskyan account of the normative ingredient in rule following – that 'right' turns out to be nothing but 'flowing from the undisturbed language organ in an undisturbed way' – can really be accepted. Is it really the sort of normativity we wanted? Do the Chomskyan meaning-facts really tell us what we *ought* to do in order to correctly apply the word "plus"? This is a tricky question since it may all too easily lead us into dubious metaphysics, into a search for superlative, 'genuinely normative' facts which really 'tell us' what we ought to do, instead of the natural, psychological facts offered by Chomsky; or for a mysterious 'logical compulsion' instead of a merely psychological one.[19]

Note, by the way, that this psychological 'compulsion' postulated by a Chomskyan theory is, strictly speaking, not a compulsion at all but merely a certain 'incitement or inclination'. So Descartes has called it and, following Descartes, also Chomsky himself. Chomsky writes: "In the Cartesian view, the 'beast machine' is 'compelled' to act in a certain way when its parts are arranged in a particular manner, but a creature with a mind is only 'incited or inclined' to do so [...]. Human action, including the use of rules of language, is free and indeterminate. Descartes believed that these matters may surpass human understanding: We may not 'have intelligence enough' to gain any real understanding of them [...]. [In fact], it is difficult to avoid the conclusion that serious problems are touched on here, perhaps impenetrable mysteries for the human mind, which is, after all, a specific biological system and not a 'universal instrument which can serve for all contingencies', as Descartes held in another context." (Chomsky 1986, p. 222) Thus, when talking about the

---

[19] Compare Wittgenstein's ironic remark in Wittgenstein 1953, § 140: "For we might also be inclined to express ourselves like this: we are at most under a psychological, not a logical, compulsion. And now it looks as if we knew of two kinds of case."

'flowing from the undisturbed language organ in an undisturbed way', one must not assume a deterministic mechanism but rather a mechanism that is compatible with what one calls our 'free will'. We are, of course, free to violate our norms.

But despite all this refinedness of Chomsky's view, it may nevertheless seem much more appropriate to try to account for the normative ingredients in meaning not by looking for psychological facts – in the Chomskyan sense of a psychology of *individuals* – but for *social* facts. After all, it is a truism that we are often corrected by others – others *tell* us what we *ought* to do – and that we are in general ready to accept their corrections (if they are able to convince us). So it is a truism that the community at least reflects our standards for 'right' and 'wrong'. In the present context, however, the question is rather whether the community also sets these standards. This is in fact claimed by many philosophers today. It is the essential characteristic of the skeptical solution with which Kripke wants to bypass the meaning skepticism developed in his book, and it is also suggested by the so-called 'nonindividualistic' considerations of Hilary Putnam and Tyler Burge,[20] which seem to demonstrate in a flash that meanings (and beliefs, and intentional states in general) 'are not in the head'. Thus, it may be the community with its shared language that we have to concentrate on when looking for the normative ingredients of meaning.

I cannot go into that here. And I need not go into it (at least at this stage of our investigation) since Chomsky maintains that it is misguided almost throughout and since it seems to me that Chomsky has a right to say that as long as the essential weaknesses of his own position have not been shown. If the latter has not been done in a convincing way, Chomsky may be justified in reacting as follows: "I say nothing about the [Kripkean] normativity requirement; nor, in my view, does anyone else, if we mean something intelligible and substantive. The discussion relies throughout on unanalyzed notions of 'community' and 'language [shared by the community]' and others that have been given no clear sense." (Chomsky 1987, p. 192) And Chomsky may furthermore point out that our behavior within the community may be explained by his own theorizing, at least partly, and that it need not be considered as a fundamental fact, as suggested by many versions of the community view (e.g., the Wittgensteinian and Kripkensteinian one). In other words, Chomsky may simply retire to his naturalistic perspective in order not to be irritated by his philosophical opponents. Thus, it seems to me, the normativity objection alone has only little force. Chomsky's model possesses

---

[20] See, e.g., Putnam's 'division of linguistic labor' in Putnam 1975 and Putnam 1988, Chap. 2, and Burge's thought experiments in Burge 1979. See also Burge's general discussion in Burge 1989.

the resources to account for a certain sort of normativity, and Chomsky may well be justified in rejecting any further or other normativity requirements.[21]

3   Chomsky's Theory as an Empirical One

In this final section, I want to develop another argumentative strategy against Chomsky's view. I will concentrate on the fact that according to Chomsky even a theory about what we *mean* should be an empirical one, quite similar to theories in physics, chemistry, and biology. Chomsky never tires of emphasizing that linguistics has no special status in comparison to other sciences.[22] This claim, however, is *prima facie* anything but plausible. Apparently, linguistics is distinguished by the peculiarity that the object of its investigation – namely, language, or the language faculty or language organ – is at the same time, at least partly,[23] its medium and instrument of investigation as well. It is true that, analogously, the object of investigation in experimental physics – namely, matter – is clearly put to use in the instruments of investigation, too. This, however, is hardly comforting since, among other things, this very fact seems to be the one which causes dreadful problems, as can be seen from the never-ending discussion about the foundations of quantum mechanics. Should one not expect that similar problems will be even more dreadful in linguistics? Chomsky, however, does not seem to regard this issue as serious.[24] Let us see what happens if one does regard it so.

---

[21] Of course, there are down-to-earth pragmatic norms like "One ought to use words as one has used them in the past if one wants to be easily understood" or "One ought to tell the truth", which, however, are not at issue here since they do not have very much to do with 'meaning-constituting facts'. See the illuminating discussion of these norms, and of the topic "normativity" in general, in Bilgrami 1992, Chap. 3, especially pp. 109–113.

[22] See, e.g., Chomsky 1986, pp. 224 and 239; Chomsky 1988, pp. 7f.; Chomsky 1992a, p. 106 (where Chomsky explicitly says that even "the study of meaning [has] the same status as any other"); and, most extensively, Chomsky 1995.

[23] Of course, our thinking involves not only the language faculty but other cognitive faculties as well.

[24] Notice that, in Chomsky's view of linguistics, the fact that the object and the instrument of investigation coincide (at least partly) takes a particularly straightforward form, since according to this view there is essentially only *one* human language, namely, the language characterized by what Chomsky calls "universal grammar", that is, the grammar underlying all human languages. Chomsky explains this by the following thought experiment: "A rational Martian scientist, studying humans as we do other organisms, would conclude that there is really only one human language, with some marginal differences that may, however, make communication impossible, not a surprising consequence of slight adjustments in a complex system. But essentially, human languages seem cast to the same mold, as one would expect, considering the most elementary properties [of the language organ] and the manner in which it develops." (Chomsky 1991, p. 4;

So let us concentrate on the fact that a Chomskyan theory about what we mean is to be an empirical one, comparable to theories in physics, chemistry, and biology. Chomsky, of course, is actually forced to present it as such because, according to his own standards, only an empirical theory is capable of presenting respectable *facts* as to what we mean by our words and signs. These facts must be able to be expressed by respectable theories. Then, however, we must recognize the possibility that such theories, and that even the best of them, may state that by "+" we do *not* mean the plus function but a quus function. Or, being a little more cautious and using the Kripkean stage-setting which distinguishes between our (non-problematized) present and our (problematized) past use of "+", we must recognize the possibility that such theories may state that in the past by "+" we did not mean the plus function but a quus function. Because of the obvious scarcity of the data available to us, we must reckon with such a possibility, however surprising or abstruse it may seem. Of course, it is just this scarcity of the data which is also the argumentative basis of Kripke's skeptic. Let us call a theory which claims that by "+" we meant the plus function "plus theory", and a theory which claims that by "+" we meant a certain quus function "quus theory". In light of the scarcity of our data, then, we have to recognise the possibility that our theories about what we meant (supposing we have got such theories at all) will be quus and not plus theories. And Chomsky, who claims that among all candidates for such theories exactly one may turn out to be the best, must recognize the possibility that even this best theory may be a quus theory. As far as I know, Chomsky never explicitly discussed a case like this. However, let us assume from now on that this case has occurred.

---

see also Chomsky 1995, p. 13) Now, with the help of this one human language, we study this one language itself. Of course, other faculties of the mind may help, too, but this does not alter the fact that we have a characteristic sort of self-reference here. It would be surprising if this self-reference did not lead to special problems.

Maybe Chomsky is bored with reflections about self-reference. Maybe he tends toward an opinion like the following, expressed by a mathematician in the recreational mathematical journal *The Mathematical Intelligencer*: "If a typical working mathematician shows an interest in [the phenomenon of self-reference], it is more on the recreational side, in a Gardner mood. This is why in *The Mathematical Intelligencer* the place occupied by self-reference is disproportionate to the importance an ordinary reader of this journal would attach to it in a working mood. This is also why in a paper on self-reference [...] one can hardly expect to learn much. One just gets the thrill of something amusing, but also whimsical, confused, and a little bit unnerving – intriguing but not worth pursuing seriously – something basically hollow. One abandons such papers without enlightenment, without any sense of achievement. Like every amusement, in excessive quantities they may lead to boredom." (Døsen 1992) This, however, is not true of mathematical logic, of course (never heard about Gödel?), and I think it is not true of linguistics either.

We should perhaps get clear, first of all, what Kripke's skeptic would say to such a situation. Kripke's skeptic would, of course, admit that the data might make us accept only quus theories (for instance, if we all had displayed a concurrent quus behavior when the numbers to be 'added' exceeded a certain limit), but (the skeptic would argue) our stock of data would still remain restricted and compatible with a host of different quus theories. In order to choose one theory among them as 'the best', we therefore must bring methodological considerations into play, for instance simplicity considerations; but (so Kripke's skeptic would continue) in the present context, which is not concerned with epistemological issues but rather with the metaphysical question about the factuality of what we meant, simplicity considerations are misplaced (see Kripke 1982, pp. 38f.).

This argument, however, seems inappropriate to me since it makes use of an excessively metaphysical, theory-transcendent notion of a 'fact'. If we look at the actual assertions concerning factuality or nonfactuality occurring in the natural sciences, we will discover that in practically all cases, even when referring to the 'data' themselves, these assertions are essentially based on methodological considerations.[25] If the skeptic declares these considerations to be out of place, he simply contradicts well-tried scientific practice and need not be taken seriously.

If I see it correctly, Kripke has not put forward any really convincing argument that would exclude the possibility to be envisaged by a Chomskyan approach that a specific quus theory might turn out to be the best theory about what we meant and that this theory might in particular present us with a specific fact as to what we meant by "+".[26]

---

[25] This standard insight of modern philosophy of science has been presented in a particularly convincing way in Friedman 1983, pp. 266–273, and Railton 1989.

[26] There is one argument of Kripke's that *may* turn out to be convincing in the end, which, however, I am no competent judge of. It is the argument that a physical organism (or a machine) doesn't possess a *unique, intrinsic* computational state and that, therefore, Chomsky's desired scientific theory, which basically would be a theory dealing with physical organisms, could not distinguish between meaning plus and meaning quus. In Kripke's own words, "a concrete physical machine, considered as an object without reference to a designer [as it would be the case of a biologically described organism], may (approximately) instantiate a number of programs that (approximately, allowing for some 'malfunctioning') extend its actual finite behavior. If the physical machine was not designed but, so to speak, 'fell from the sky', there can be no fact of the matter as to which program it 'really' instantiates." (Kripke 1982, p. 39, footnote 25) What remains unclear to me here is how precisely we are to understand the notion of a 'machine'. It seems at least conceivable that when using a sufficiently differentiated notion of a machine, we might be justified, pace Kripke, in claiming that machines can have unique, intrinsic computational states. Chomsky himself, though in a different context, has expressed his dislike, as a theoretician, of the notion of a machine: "the concept of 'machine' has no clear sense. [...] We have no *a priori* concept of what constitutes 'the material world' or 'body' – hence 'machine'." (Chomsky

Kripke's considerations, however, can be modified in such a way that convincing arguments may emerge. To show this, I must explain the so-called 'skeptical paradox' developed by Kripke, which I have not mentioned until now. This Kripkean paradox arises in the following way: As repeatedly stressed, in the Kripkean setting the meanings of our *present* utterances are considered as uniquely determined and are taken for granted. Seen from this meta-level of our present utterances, the object-level of our *previous* utterances, then, appears indeterminate with respect to the meaning of these utterances. But there is *no relevant difference* between meta- and object-level with respect to meaning. Hence, with respect to meaning, these two levels may be identified, and we immediately obtain a contradiction between determinacy and indeterminacy. This is Kripke's skeptical paradox.

A structurally similar contradiction is threatening now in the Chomskyan quus theory. It runs as follows: On the meta-level of our scientific study of human beings, we seem to presuppose, when using mathematics (which we certainly have to do), that by the plus sign we mean the plus function. Seen from this meta-level, however, the objects of our investigation, the human beings in the past, now appear to have meant a quus function, and not the plus function. But since the subject of investigation, i.e. the investigator, can be identical with his object, and since there is no relevant difference between this subject in the past and this subject in the present, we immediately obtain a contradiction between the presupposition of meaning plus and the result of meaning quus by the plus sign.

To put it more concisely: Since Chomsky must consider it as a possible result of his research that what we in fact mean by our words is different from what we believe them (presuppose, take for granted) to mean, he must reckon with the possibility that what *he himself* means is different from what, during his research, he believed them (presupposed, took for granted) to mean.

Thus, Chomsky ends up in a strange, contradictory situation. He seems to be running into just the sort of predicament that has been so vividly described in Kripke's book: that in the end the rug is pulled out from under our own feet. For, we formulated the Chomskyan quus theory in terms whose meanings we took for granted; and the theory then implies (or seems to imply) that we do not mean what we took for granted. From that very moment we can no longer trust our whole language. So, we cannot say, for instance, "Our theory has shown that by the word 'plus' we mean quus and not plus", since in this statement we still seem to presuppose that by the word "plus" we mean plus (and that the meaning of the word "quus" has somehow to be explained with the help of the word "plus"), contrary to what the statement itself seems to say. Consequently, we have no good reason at all to say, "By

---

1987, pp. 191f.) This, however, if directed against the Kripkean point (which has not actually been done by Chomsky), would be too sloppy a reaction, and the matter surely deserves careful investigation. What we need is not an *a priori* concept of 'machine' but an empirically fruitful one.

the word 'plus' we mean quus and not plus", since this result belongs to an inconsistent system of thought. Meaningful speech seems to break down.

Let us discuss the situation which Chomsky would run into with a quus theory in a little more detail. I will now ignore the Kripkean distinction between what we meant in the past and what we mean at present: It was only a ladder which we used to climb up to our preceding result, and this ladder can now be thrown away. One should distinguish, then, between the following two cases. Case (a): Chomsky's theory says that our meaning quus by the plus sign already concerns numbers which we can feasibly write down and calculate with. Case (b): Chomsky's theory says that our meaning quus by the plus sign only concerns numbers bigger than those which we can feasibly write down and calculate with.

With regard to case (a), let us suppose, for example, that Chomsky's quus theory says that according to what we mean by the plus sign, we should say "68 + 57 = 5" and not "= 125". We can test this theory, then, by actually calculating "68 + 57". (Presuppose that we never calculated "68 + 57" before.) Suppose that we in fact calculate "68 + 57 = 5", that is, that we in fact *say* or *write* "68 + 57 = 5". If we take this actual saying or writing to be a criterion of what we mean by the plus sign, then Chomsky's theory is confirmed. The question, however, arises of whether this may be considered a case where what we believe we mean is different from what we in fact mean. Does it make sense to say that I believed I meant the plus function, but that, one fine day, by finding myself actually writing down "68 + 57 = 5", I discovered, to my own surprise, that I did *not* mean it? To my mind, this is plain nonsense.[27] What *does* make sense is that I would deem my writing "68 + 57 = 5" a mistake and would correct it in order to get into agreement with what I really meant. But this, then, is no longer a case where Chomsky's quus theory is confirmed, and it is not a case where what I believe I mean differs from what I mean.

Certainly, the sort of self-correction just described belongs to our ordinary behavior and, moreover, it need not involve a community. It is a plain fact that we often correct ourselves without thereby relying on the judgments of others. Thus, in referring to this behavior, we do not leave the individualistic perspective of Chomsky's theory and, I think, we also need not exceed the weak sort of normativity accepted by Chomsky (by means of his competence–performance distinction). Chomsky may very well have a place for this behavior in his naturalistic model of rule following (see Section 2). But it now seems that Chomsky's model, when taken as confirmed by our actual behavior, including the behavior of correcting mistakes, cannot show us that what we mean differs from what we believed to mean.

Of course, if I am *not* surprised when finding myself writing "68 + 57 = 5" and if I have *no* tendency to correct this equation, the claiming of such a

---

[27] It seems to me that this kind of nonsense would deserve further study, which, however, cannot be pursued here.

difference would appear out of place anyway: the plus sign could then simply be regarded as meaning a function which assigns the value 5 to the argument (68,57) and not the value 125 – and everything would be in order. Maybe, against this a Chomskyan might try to argue that our actual behavior is not at all a criterion of what we mean: the question as to what we mean is a question concerning our competence, and our competence has to be sharply distinguished from what we actually do, that is, from our performance. So, a Chomskyan might say, for example, that our actual *use* of the plus sign might very well be in accordance with the plus function, whereas what we *mean* by the plus sign is nevertheless a quus function. He could present us a model of our mind, where a certain distinguished part of the mind – what Chomsky calls a "module of the mind" – which is considered to be responsible for what we mean, represents a quus function, whereas, because of certain 'disturbing factors', what we actually do accords with the plus function. This, however, certainly would be a dubious reply (and I do not in fact charge Chomsky himself with it). For, we can now ask why we should consider only the restricted module, representing the quus function, to be the 'meaning module' and why it is not the whole complex consisting of this module *and* the so called 'disturbing factors'. After all, it is this whole complex which is responsible for our use of language. I am sure the reasons for considering this latter complex as representing what we mean will always be the more convincing ones.

If the preceding considerations are correct, then we must conclude that with respect to numbers which we can feasibly write down and calculate with, Chomsky's theory *cannot* – in a very strong sense of "cannot" – contradict our normal rule-following behavior, including, of course, the important behavior of correcting mistakes. Maybe Chomsky will be quite content with this result, but to my mind it raises doubts concerning Chomsky's scientistic outlook which tends to play down our actual behavior ("performance"). Furthermore, it may seem, then, that the really strong point Chomsky has to make against opponents who, like Quine, tie 'meaning' to behavior and to nothing else, must essentially rely on case (b), the case in which Chomsky's theory says that our meaning quus by the plus sign concerns only numbers bigger than those which we can feasibly write down and calculate with. In this case, Chomsky's quus theory tells us what we *would* do if some of our limitations – limitations in time and memory, say – and other 'disturbances' were absent, and it tells us that we would use the plus sign in a quus-like manner.

What's problematic about that? One may argue that we would have discovered, then, that in the counterfactual situation, where we 'add' in the quus-like manner, we would add *wrong*; that we would still mean plus by the plus sign, but we would simply miscalculate. I doubt, however, that this by itself is a good reply to Chomsky. Suppose that Chomsky's theory, which tells us that in certain counterfactual situations we would use the plus sign in a quus-like manner, is empirically well-confirmed. What could be our reasons,

then, for claiming that it is *not* a theory about what we mean by the plus
sign but rather a theory which predicts certain miscalculations? Where is
our standard which determines what miscalculations are and are not? This
standard is not to be seen at all, and Chomsky may easily maintain, I think,
that his theory is a theory about what we mean.

Chomsky can maintain this against the meagre reply just mentioned. But
what if we back up this reply by my main argument which says that he
is running into a contradiction? When he constructed his theory, didn't he
thereby presuppose – contrary to what the theory then claims – that by the
plus sign he meant the plus function? In a sense he surely did presuppose it,
and if we take this presupposition at face value the contradiction cannot be
disputed. And we *have*, of course, a strong, almost irresistable tendency to
take this presupposition at face value. It hardly seems possible to dispel the
impression that we *know* what we mean by "plus" and that what we mean
by it is without doubt the plus function and nothing else. That we assume
this authoritative *self-knowledge* seems to belong to our self-description as
thinking subjects which we simply are not ready to give up and which we
take for granted in our intellectual endeavors. At points like this, Chomsky's
strategy is to say that one should *separate* our ordinary self-descriptions from
the scientific theorizing about the human mind (see Chomsky 1992b and
1995). But this separation doesn't seem to be possible in the present case
where assertions like "I know what I mean by 'plus'" seem to be important
for the scientific theorizing itself. They reflect the fact that we all take our
arithmetic for granted and that we have no sensible idea as to what it would
be like not to take it for granted. The arithmetic notion *plus* is both a common
sense *and* a scientific notion, and our common sense understanding of it flows
directly into our scientific inquiries.[28]

A cognitive scientist may ask what this supposed self-knowledge really
consists of; but I think he will have a hard time convincing us that it might
be an illusion (or a kind of self-deception as in Freudian cases). Of course,
the Kripkean skeptic even tries to convince us that it in fact *is* an illusion;
however, he has no chance of succeeding since, if nothing else helps, we are
ready to silence him by a skeptical solution (i.e., a solution that does without
the factuality of meaning) which respects our ordinary self-descriptions. And
I think in the Chomskyan case, too, if nothing else helps it may be some sort
of skeptical solution which should be adopted in the end (see Sect. 1).

Obviously, the potential contradiction which Chomsky is running into is
a very simple one and of a very general nature. It arises in any situation in
which, when scientifically theorizing about certain cognitive capacities, we

---

[28] Needless to say, my appeal in this paragraph to our self-knowledge is only a leap
in the dark. It only points in the direction of further investigations which are
necessary at this point; i.e., it only locates my argument within the philosophical
discussion about the nature of self-knowledge, a discussion which I cannot go into
here.

cannot avoid putting these same capacities to use and presupposing that we thereby know what we are putting to use. Our theory, then, *must not* contradict this supposed knowledge; but at the same time, by being an empirical one, it should *be able* to contradict it. One may suspect that this argument against Chomsky is too general to have much force. However, I doubt that this is true, since the argument touches exactly that point – a point concerning *self-reference* when theorizing about ourselves – where one should actually expect a sore spot of cognitive science. Of course, what I have said so far is only the beginning of a serious examination of this spot. But I think it worth examining in every detail.

References

Bilgrami, Akeel (1992): *Belief and Meaning* (Blackwell, Oxford & Cambridge, Mass.)

Boghossian, Paul A. (1989): "The Rule-Following Considerations", *Mind 98*, pp. 507–549

Burge, Tyler (1979): "Individualism and the Mental", in: French/Uehling/Wettstein 1979, pp. 73–121

Burge, Tyler (1989): "Wherein is Language Social?", in George 1989, pp. 175–191

Chomsky, Noam (1982): *The Generative Enterprise* (Foris Publications, Dordrecht)

Chomsky, Noam (1986): *Knowledge of Language* (Praeger, New York)

Chomsky, Noam (1987): "Reply [to Alexander George and Michael Brody]", *Mind & Language 2*, pp. 178–197

Chomsky, Noam (1988): *Language and Problems of Knowledge* (MIT Press, Cambridge, Mass.)

Chomsky, Noam (1991): "Language from an Internalist Perspective", ms.

Chomsky, Noam (1992a): "Language and Interpretation: Philosophical Reflections and Empirical Inquiry", in: Earman 1992, pp. 99–128

Chomsky, Noam (1992b) "Explaining Language Use", *Philosophical Topics 20:1*, pp. 205–231

Chomsky, Noam (1995): "Language and Nature", *Mind 104*, pp. 1–61

Døsen, K. (1992): "One More Reference on Self-Reference", *The Mathematical Intelligencer, Vol. 14, No. 4*, pp. 4f

Earman, John, ed. (1992): *Inference, Explanation, and Other Frustrations* (University of California Press, Berkeley and Los Angeles)

French, P., Uehling, T. and Wettstein, H., eds. (1979): *Midwest Studies in Philosophy 4: Studies in Metaphysics* (University of Minnesota Press, Minneapolis)

Friedman, Michael (1983): *Foundations of Space-Time Theories* (Princeton University Press, Princeton)

George, Alexander, ed. (1989): *Reflections on Chomsky* (Blackwell, Oxford)

Horwich, Paul (1995): "Meaning, Use and Truth", *Mind 104*, pp. 355–368

Kitcher, Philip and Salmon, Wesley C., eds. (1989): *Scientific Explanation (Minnesota Studies in the Philosophy of Science XIII)* (University of Minnesota Press, Minneapolis)

Kripke, Saul (1982): *Wittgenstein on Rules and Private Language* (Blackwell, Oxford)

Meggle, Georg and Wessels, Ulla, eds. (1994): *Analyomen 1* (Walter de Gruyter, Berlin)

Mühlhölzer, Felix and Emödy, Marianne (1994): "Über eine mögliche Inkonsistenz in Chomskys Auffassung von Sprachregeln", in: Meggle/Wessels 1994, pp. 481–492

Mühlhölzer, Felix (1995): "Science Without Reference?", *Erkenntnis 42*, pp. 203–222

Putnam, Hilary (1975): "The Meaning of 'Meaning'", repr. in H. Putnam, *Philosophical Papers, Vol. 2* (Cambridge University Press, Cambridge)

Putnam, Hilary (1988): *Representation and Reality* MIT Press, Cambridge, Mass.)

Quine, Willard Van (1961): "Notes on the Theory of Reference", in W.V. Quine, *From a Logical Point of View*, 2nd Ed. (Harper & Row, New York) pp. 130–138

Quine, Willard Van (1969): "Ontological Relativity", in W.V. Quine, *Ontological Relativity and Other Essays* (Columbia University Press, New York) pp. 26–68

Quine, Willard Van (1981): *Theories and Things* (Harvard University Press, Cambridge, Mass.)

Quine, Willard Van (1992): *Pursuit of Truth*, Revised Ed. (Harvard University Press, Cambridge, Mass.)

Railton, Peter (1989): "Explanation and Metaphysical Controversy", in: Kitcher/Salmon 1989, pp. 220–252

Wittgenstein, Ludwig (1953): *Philosophical Investigations* (Blackwell, Cambridge)

Wright, Crispin (1989): "Wittgenstein's Rule-following Considerations and the Central Project of Theoretical Linguistics", in: George 1989, pp. 233–264

Frege's Sense and Tarski's Undefinability*

Volker Beeh

Heinrich-Heine-Universität Düsseldorf, Universitätsstr. 1,
D-40225 Düsseldorf, Germany. e-mail: beeh@mail.rz.uni-duesseldorf.de

Obviously many expressions designate the same. Under certain conditions the demonstrative pronoun *this* refers to the picture denoted elsewhere by *the photograph of my mother* and the relative clause *whoever discovered the elliptic form of the planetary orbits* refers invariably to the man known as *Kepler*. However it is less evident why complex expressions referring to the same things are not mere luxury like synonymous words.

1.
    a) The evening star is the same as the morning star.
    b) Kepler is the man who discovered the elliptic form of the planetary orbits.
    c) The intersection of a and b is identical with the intersection of b and c.
    d) $2 + 3 = 5$.

Equations and inequations state identity or difference of the things and play an important role both in ordinary and scientific discourse. In his celebrated *treatise [o]n Sense and Reference)* Gottlob Frege presented an analysis which has been debated for one hundred years. If the expressions *evening star* and *morning star* differ only superficially, i.e., agree in all respects except their strings of letters or sounds, then the equation *the evening star is the morning star* does not say more than *the evening star is the evening star*, i.e., nothing. The fact that the former expresses a major discovery of ancient astronomy cannot be accounted for by the different sounds or characters of *evening star* and *morning star* but presupposes a difference in their senses. Otherwise, it must have been equally interesting whether humanity is mankind. The assumption that certain expressions differ in sense and nevertheless agree in reference has never been seriously challenged at least in its naive form and will be accepted in all what follows. To do Frege philological justice one has to add that he distinguished between German *Sinn* and *Bedeutung*. Whereas the former is the natural choice and corresponds to English *sense* the latter even in contemporary German is misleading. Frege had in mind what elsewhere is called *reference, denotation* or *designation*, but certainly not *meaning*. In

---

* A contribution with similar content will appear in German as *Verständnis ohne Erkenntnis* in Jochen Lechner (ed.), *Analyse, Rekonstruktion, Kritik*. Frankfurt a.M. 1997. I must offer specific thanks to Joachim Bromand (Düsseldorf) who gave me helpful comments on earlier versions of this paper.

what follows *sense* and *reference* translate *Sinn* and *Bedeutung*, respectively. But, Frege not only maintained that the cognitive value of equations presupposes a difference both in their terms and in the senses of their terms, he suggested at least that, conversely, the difference in senses explains the cognitive value of equations. Unclear as it is, the notion of cognitive value inspired much of the interest in Frege's distinction. The main purpose of this paper is to demonstrate the failure of this latter idea. The cognitive values of sentences emerge somewhere else and appear in better resolution through the negative or limitative results of logic and arithmetic, e.g., through Tarski's theorem of undefinability of truth.

1. Frege's Problem

In the opening and closing passages of *Sense and Reference* Frege argued in the format of (2):

2.
  a) We understand (well-formed) equations.
  b) Instances of the schema **a=a** are generally trivial. In contrast many statements of the form **a=b** do have cognitive value.
  c) Various expressions refer to things in different ways. The mode of presentation belongs to their sense (Art des Gegebenseins).
  d) Understanding terms is grasping their senses, grasping the senses does not include recognizing their references.
  e) The cognitive value of **a=b** rests on the difference in the senses of the terms **a** and **b**.

The following concentrates on equality and difference and later covers other relations, too. Things equal themselves and differ from all other things. Equality and difference are so to speak trivial relations. We are not interested in them as such but in truth and falsity of equations and inequations, i.e., of sentences. Inequations negate equations and difference negates identity and this is why the following may disregard the negative counterparts. I will refer to sentences like $a=a$ which hold a priori and are referred to by Frege and Kant as analytic simply as *trivial*. In such cases difficulty of comprehension reaches a minimum, which ordinary language deprives of the quality of understanding in the same sense in which it seems averse to subsume air under gases. Equations of the form **a=b** do not always hold a priori. However I do not call them *a posteriori* or *synthetic*, but simply *non-trivial* or *difficult*. Of course, not all sentences of the form **a=b** are difficult and distinctions are gradual.

There is a tendency to assume that Frege pursued the establishment of senses. In contrast to the title of his treatise I follow Gregory Currie[1] and oth-

---

[1] Gregory Currie quotes an illuminating passage of a letter from Frege to Peano written in 1896: "...the whole content of arithmetic ... would be nothing more

ers and surmise that the introduction and conclusion of *Sense and Reference* do not only form a rhetorical frame, but indicate Frege's original and real problem, i.e., the distinction between trivial and difficult propositions. Because triviality is so to speak trivial in itself the problem reduces to question (3).

3.

    a) What is comprehension of equations without cognition of their truth?

    b) What can obstruct cognition of truth in understanding of equations?

These questions may have inspired Frege to his distinction between sense and reference. Later he sought independent evidence for the necessity of senses and believed to have found it in his analysis of indirect speech *(oratio obliqua)*.

Some comments on Frege's examples will follow. By scheme **a=b** he understood firstly equations connecting proper names in a narrower sense, like *Aristotle*. But, equations like *the Ateb is the Alpha* sound somewhat artificial. There is an enormous amount of scholarly literature on proper names, but because languages use only a restricted number of them, their importance must not be exaggerated. Secondly, propositions like *the evening star is the morning star* appear more natural. Their terms are compressed descriptions which are interesting from the perspective of word formation. But again languages use only a restricted number of lexicalized abbreviations. For reasons of simplicity Frege did not separate proper names from complex terms in *Sense and Reference*.[2] Thirdly, the sentence *the intersection of a and b is identical with the intersection of b and c* appears to demonstrate the point best, because the complexity of the terms evokes to some degree a true difficulty. The number of complex terms is unlimited, and this is why this example indicates the real problem calling for an explanation. Frege in his attempt to simplify his argument apparently went too far by using proper names.

Frege discussed equations of natural languages and of geometry. Can we not look upon terms of arithmetic like **5 + 1** and **2 × 3** as designating the same number in different ways and as expressing different senses? At least in *Sense and Reference* Frege side-stepped with respect to his project to reduce

---

than boring instances of this boring principle [the principle of identity]. If this were true mathematics would indeed have a very meager content. But the situation is surely somewhat different. When early astronomers recognized that the evening star [Hesperus] was identical with the morning star [Lucifer] ... this recognition was incomparably more valuable than a mere instance of the principle of identity ... even though it too is not more than a recognition of identity. Accordingly, if the proposition that 233 + 798 is the same number as 1031 has greater cognitive value than an instance of the principle of identity, this does not prevent us from taking the equals sign in '133+798=1031' as a sign of identity." (Currie, p. 102, *Wissenschaftlicher Briefwechsel*, p. 195)

[2] *On Sense and Reference*, p. 24

arithmetic to pure logic. And for logic Frege at least later considered the possibility of relating identity of senses to logical equivalence.

4.

> In order to decide whether sentence A expresses the same thought as sentence B, only the following means seem possible to me [...]. If both the assumption that the content of A is false and that of B is true and the assumption that the content of A is true and that of B is false lead to logical contradiction [...] without using sentences not purely logical, then nothing can belong to a content of A which does not belong to the content of B [...]. Are there other means to decide [...] whether two sentences express the same thought? I don't think so.[3]

According to this criterion all arithmetical truths express the same sense provided arithmetic is reducible to pure logic. In this case it would be not necessary nor possible to explain the cognitive value of arithmetical equations by means of their thoughts and their thoughts in turn by means of the senses of their terms. After reduction of arithmetic to logic proved impossible the difference of terms like $5 + 1$ und 2×3 in sense must be accepted.[4] But is logical equivalence a serious candidate for a criterion of sense-identity? Whereas I do not see any reason to deny different senses even to logical equivalents like $\neg P \vee Q$ and $P \rightarrow Q$ this question would lead us too far afield.[5]

2 Understanding Equations

The opening passages of *Sense and Reference* demonstrated that in the framework of naive semantics this problem leads to an absurd situation and takes refuge in senses. If comprehension of equations like **a=b** includes cognition of the references of **a** and **b** it inevitably includes cognition of truth. If we choose a clear notion of knowledge it seems not possible to know things without knowing their identity or difference. But mind does not work like this. In many cases we seem to understand equations without knowing about their truth value. Where the objects are out of reach, our understanding must be content with senses. Whatever senses may be, their function as a substitute for things is undebated. Frege answered question (3a) by stating that attempts to understand equations from time to time do not reach their aim.

---

[3] *Wissenschaftlicher Briefwechsel*, p. 105f. (Letter from Frege to Husserl, 9.12.1906. translation by V.B.)

[4] See note 2!

[5] Andreas Kemmerling stated correctly (p.10) that in the case of logically equivalent sentences identity of senses ("Äquipotenzkriterium", p. 5) is not compatible with compositionality of senses ("Teil-Kriterium", p. 9), but did not decide against equivalence.

In the case of equations connecting proper names in the narrow sense as in the *Ateb is the Alpha* or lexicalized abbreviations like in *the evening star is the morning star* there is a temptation to think in contrast that absurdity accounts for the correct analysis. It is to be expected that Frege's analysis gains in significance in difficult equations. Unfortunately here the same absurdity arises against the background of Frege's theory of functions. Consider again the equation *the intersection of a and b is the same as the intersection of b and c* and let us write for short $a \times b = b \times c$. Let the letters a, b and c designate the medians d, e, and f of some triangle, let the cross express the function ϕ, which matches lines to their intersections, and let the equals sign express the characteristic function $\chi_=$ of identity. Now if the functions ϕ and $\chi_=$ were ordinary things like d, e, and f, semantic interpretation of the equation $a \times b = b \times c$ would create a (redundant) list $\langle d, \phi, e, \chi_=, e, \phi, f \rangle$.[6] The five things ϕ, $\chi_=$, d, e, f do not feel urged to react to each other and to change into a truth value.

The reference of a well-formed sequence of symbols is not the sequence of the references of the symbols. We are acquainted with the convention that $\langle d, \phi, e, \chi_=, e, \phi, f \rangle$ as a notational variant of $\chi_=(\phi(d, e), \phi(e, f))$ counts as the result of the application of the functions ϕ and $\chi_=$ to their arguments d, e, and f, respectively. The notion of an *application* does not fit into Frege's conception of functions. Moreover, conventions are arbitrary and both facts may have persuaded Frege to reject this approach. Frege assigned the task of this convention to functions themselves and modeled them in such a way that they – combined with their arguments – develop without logical phases into the respective function values, or better: that combined with their arguments they *are* values. In this image functions prey on things and with saturation develop into what they eat. In Frege's theory of functions the interpretation of the sentence $a \times b = b \times c$ yields $\langle d, \phi, e, \chi_=, e, \phi, f \rangle$, i.e., a notational variant of $\chi_=(\phi(d, e), \phi(e, f))$ which in turn changes into a truth value without running through intermediate logical stages. This picture must not prevent us from seeing the difficulties of understanding where we are used to experience them, i.e., on the way from perception of equations to cognition of their truth values. Frege's model of functions does not allow for obstacles along the way and renders all understanding equally trivial. It does not explain partial understanding or difficulties.

Frege did not create an artificial problem. The absurdity shows up as well in Tarski's semantics.

5.

The mastery of a syntax, a semantic interpretation S11 of the symbols and a definition S2 of truth cannot explain the difficulties of equations:

---

[6] Here and later I neglect the internal structure or bracketing of sequences like $\langle d, \phi, e, \chi_=, e, \phi, f \rangle$.

S1 $S[a]=d$, $S[b]=e$ and $S[=] = \chi_=$.
S2 $a=b$ is true in S if and only if $S[=](S[a], S[b]) = \chi_=(d, e) = 1$.

The following line of argument corresponds to Frege. Application of rules S1&2 to the equation $a=b$ includes interpretation of the symbols a, $=$, and b, formation of $S[=](S[a], S[b]) = \chi_=(d, e)$, and determination of the truth value. If understanding is mastery of S1&2 there is no room for difficulties and understanding leads without any obstacles to a decision about truth. Tarski's semantics sees equations from the perspective of a metatheory which makes them equally trivial. Partial or difficult understanding is not possible.

The success of this semantic theory hides its absurdity. Assume $a=b$ is a really difficult equation, i.e. not an example with secretly known solution, and let us try to analyse it in the framework of (5). If $a=b$ is really difficult, where does the semanticist get its solution? And if the semanticist knows its solution, why should it have been difficult? If an equation is difficult its semantics cannot be simple. In general the analysis of trivial equations is trivial and the analysis of difficult equations is difficult. The difficulty of semantics mirrors the difficulty of its sentences – at least if we have in mind what I will call complete semantics.

3 Division of the Semantics of Words

The introduction of senses amounts to a division of the semantics of words or symbols into an easy first stage and a difficult final stage. As a consequence this leads to the concentration of understanding in the former and to the concentration of cognition and decision of truth in the latter. Frege dubbed the objects *senses* and *references*, respectively.

6.

Frege split up the interpretation S into an assignment I of senses to expressions and an assignment E of references to senses. He explained understanding with I, cognition of things and truth-values with E and difficulty of sentences with the difference between I and E.
$I[a] = \alpha$ and $I[b] = \beta$
$E[\alpha] = S[a] = d$ and $E[\beta] = S[b] = e$

It goes without saying that senses should differ from references and that different expressions agree in senses only exceptionally.

The notions of *comprehension* and *grasping* senses seem unstable in ordinary language. We understand expressions, things, facts, and combinations of them. On the one hand we say that a lack of understanding of the underlying facts keeps us from solving an equation. This variant of comprehension aims at full truth or falsehood and inaccessibility of either of them results in incomplete understanding. On the other hand comprehension is often content with the senses of words. Frege decided for the second, counted grasping of senses as part of the ordinary mastering of language, and dissociated the transition from senses to things or truth-values from comprehension. His argument that otherwise it would be impossible to ask questions is a sound one. Indeed, if it were not possible to say something, the truth of which is not included in comprehension, general communication would be superfluous. This model renders it possible to understand completely without reaching cognition. Whereas the notions admittedly occur in the way Frege used them his phraseology is somewhat arbitrary. There is no clear borderline separating grasping of senses and cognition of things. The distinction varies from problem to problem, from speaker to speaker, and from moment to moment during mental work. But a borderline, which is everywhere and always changing and vague, provides a weak basis for the modeling of constant senses as well as for a division of semantics in Frege's or Tarski's formats.

What is Frege's division of semantics of words and model of senses designed to do – except for reminding us that we sometimes fail to understand sentences completely? Of course, sentences of the form $a=a$ are trivial because of the identity of terms. Consider equations like $a=b$ and suppose comprehension includes grasping the senses $I[a]$ and $I[b]$. If these are equal – as presumably in *humanity is mankind* – the sentences $a=b$ are again trivial and of little or no cognitive value. If they differ – as in *the evening star is the morning star* – the equations $a=b$ do not necessarily have a proper cognitive value. In fact many contemporary speakers do not feel the need to prove this sentence. Instead they learned the terms as closely related ones. Sentences like the *intersection of a and b is the intersection of b and c* or $a=b$ for short turn out to be both understandable and of some cognitive value. Comprehension may involve senses and cognitive value may involve both interest in knowledge and difficulty of proof. But, given the simplicity of the assignment I of senses to expressions the proof of $a=b$ is as difficult as that of $EI[a] = EI[b]$ and therefore as the assignment E of things to senses. With trivial E the sentences $a=b$ are trivial, with difficult E they gain in cognitive value which does not depend on the difference of senses but on the assignment E of things to senses. Frege seems to have modeled his senses by mixing up the easy initial part of the way to cognition with the difficulties on the final part. A simple division of semantics will not meet the hopes expressed in the closing passage of *Sense and Reference* where senses are declared responsible for cognitive values.

7.

> When we found 'a=a' and 'a=b' to have different cognitive values, the explanation is that for the purpose of knowledge, the sense of the sentence, viz. the thought expressed by it, is no less relevant than its reference, i.e. its truth value. If now a=b, then indeed the reference of 'b' is the same as that of 'a', and hence the truth value of 'a=b' is the same as that of 'a=a'. In spite of this, the sense of 'b' may differ from that of 'a', and thereby the sense expressed in 'a=b' differs from

that of 'a=a'. In that case the two sentences do not have the same cognitive value. If we understand by "judgement" the advance from the thought to its truth value [...] we can also say that the judgments are different. [7]

The value of money results from what you do not possess, what you need, and what you can buy with it. Cognitive values of equations result from what you do not know, what you want to know, and from the utility of senses. My central thesis contends that difference of senses is a necessary but not sufficient condition for constitution of cognitive values. Whereas Frege admitted in the first section of *Sense and Reference* that difference of senses does not always render equations difficult he at least suggested in the last section that it does. How is it possible according to Frege to understand two terms **a** and **b** without knowing if the things are equal? Because understanding is grasping of senses and grasping of senses is not cognition of things. My objection repeats this argument: How is it possible to grasp two senses without knowing if they are the senses of the same thing? Does the solution of this riddle call for supersenses? Frege characterized senses as *modes of presentation* of things.[8] How is it possible to understand a mode of presentation without knowing which thing it presents? Frege could not have meant *logical form*, e.g., that names signify things, predicate symbols express characteristic functions of predicates, connection symbols express truth functions, etc. In addition, the *mode* of presentation must contain a kind of guide to the object. If, moreover, the mode of presentation has the nature of expressions – which Frege suggests in his famous footnote – the sense is explained with the sense of an explanation. In this case Frege's senses have the status of signs and raise the question they are designed to answer.

There have been several attempts to escape from this dilemma. A somewhat different division is favored by Donald Davidson and his followers who utilize Tarski's semantic interpretation as a kind of translation into the metalanguage and who translate difficult equations **a**=**b** into not less difficult metalinguistic truth conditions of the form **S**[**a**]=**S**[**b**]. Truth conditions share their difficulties with Frege's senses. But conservation of difficulties is no explanation. The architects of the semantics of possible worlds proposed to model senses or intensions as assignments of things to possible worlds, situations or circumstances. This is based on the fact that the sense of an expression like *the next mayor* enables us to determine a person at a certain place and a certain time. Many expressions accept the world as an equal partner in language games. Hilary Putnam discovered two previously unnoticed ways in which we utilize the world in speaking.[9] In this perspective the difficulties of sentences lie in the lack of information, i.e., the freedom of one

---

[7] *On Sense and Reference*, p. 50, Moore 1993, p. 42

[8] *On Sense and Reference*, p. 26 "die Art des Gegebenseins" Geach and Black translate "the mode of presentation" (Moore p. 24)

[9] Hilary Putnam 1973

or more variables. The equation $\mathbf{a}=\mathbf{b}$ then turns out to have the deep structure $(\mathbf{a}=\mathbf{b})(\mathbf{w})$ where $\mathbf{w}$ is a variable for worlds, situations, or circumstances. Whereas this of course is a substantial aspect of language use, I do not think that it supplies an answer to Frege's question. The difficulty he indicated in his geometric example certainly does not result from freedom of some parameter. In contrast Frege used in his treatise only sentences which are more or less independent from situational knowledge. Objections of similar kind could be made against Meinong's notion of *possible things*. The distinction between actual and possible things is closely related to Frege's distinction between things and senses. In all cases a distinction bears the burden of explaining fragmentary comprehension. In all cases I raise the same objection. The use of ersatz coffee does not explain the shortage of real coffee and you cannot eat money. (I will not comment on the transformation of this question in Russell's theory of descriptions.)

4   Generalization

Up to now the argument followed *Sense and Reference* and was restricted to equations and inequations. Consider a predication like *Kepler died in misery* or in symbols $\mathbf{Pa}$, let the interpretation $S[\mathbf{P}]$ of the predicate symbol $\mathbf{P}$ be the characteristic function of the set of all people who died in misery and let the interpretation $S[\mathbf{a}]$ of the name $\mathbf{a}$ be Kepler. If comprehension of $\mathbf{Pa}$ presupposes the knowledge of the interpretations $S[\mathbf{P}]$ and $S[\mathbf{a}]$ the sentence $\mathbf{Pa}$ must be true or false, but in both cases inevitably trivial. Whoever knows Kepler and knows what it means – in a certain sense of the word – to die in misery cannot be uncertain about the truth of $\mathbf{Pa}$ or *Kepler died in misery*.[10] The problem of non-triviality is not restricted to equations and figures in all sentences. Necessity of senses for terms is so obvious in the case of equations because identity seems to contribute less to the senses of equations than terms.

Non-triviality turns out to be a serious problem. Undecidability of elementary logic points into the same direction. Whereas the totality of logical truths and falsehoods in elementary logic (predicate or quantificational logic of first order) is recursively enumerable, the set of elementary contingencies cannot be recursively enumerated (and this is why both sets cannot be recursively decided). Logical truths and falsehoods can be recognized better than contingencies in elementary logic. Here we are faced with the same problem in a different setting.

This contrasts remarkably with ordinary language. Many sentences which figure in logic as tautological, contradictory, or logically true or false occur in ordinary life but inevitably undergo contingent reinterpretation.

---

[10] Introduction of partial knowledge of things seems natural but amounts to another division of semantics which is at least as dark as the use of senses.

8.

> War is war.
> Unauthorized persons not permitted.
> If nobody goes, neither do I.
> This is good or not good.
> She does it by herself.
> Children are not children.
> This is good and not good.
> She does not do it by herself.

Ordinary discourse upgrades worthless expressions.

5   Tarskian Senses

The outcome of these considerations is the futility of all attempts to explain non-triviality of sentences by means of a voluntary division of semantics of words or symbols of some sort or other. It is mistaken to blame words or their senses for being obstacles in comprehension. Difficulty emerges in calculation of references of complex expressions, i.e., in logical or linguistic complexity. Impressed by his theory of easy saturation of functions Frege did not come to this diagnosis. Yet, this doctrine has not found general acceptance. The theory of recursive functions developed several measures of complexity of computation and decision of sentences. Here non-triviality is caused mainly by quantifiers, in particular by the interplay of several quantifiers. It is revealing to see that quantifiers are the only symbols which did not get a functional interpretation in Frege and which did not contribute to his attack on non-triviality. Frege put into the senses of terms that which ought to have been put into the complexity of alternating quantifier prefixes.

Moreover, Frege demonstrated his distinction with words like *evening star* or *morning star*. Word formations in fact abbreviate complex descriptions like *the single star shining in the evening in a certain position* and *the single star shining in the morning in a certain position*. The difficulties which are condensed into the senses of words or symbols are hardly accessible for later analysis. The study of recursively noncomputable functions should not begin with noncomputable basic functions. Instead it has to start with functions computable in the intuitive sense, and furthermore it is to synthesize noncomputability, e.g. by diagonalization. In an attempt to imitate this procedure, let us consider an artificially simple language which uses only easy words, but which generates difficult terms and sentences and let us ask two questions: What can be identified with Frege's senses and what may be blamed for difficulties of sentences? The best choice is the language of elementary arithmetic (first order arithmetic) the intended interpretation of which is more or less trivial in contrast to the notion of arithmetical truth. According to Tarski's theorem of 1953 the totality of true sentences of arithmetic cannot be defined in arithmetic nor can it be recursively decided. Arithmetic creates difficult sentences out of easy symbols literally before our eyes.

The interpretation **S** is conceived in such a way that it provides the arithmetical symbols **0**, **S**, **+**, ×, as well as all symbols of logic including the equals sign = with their usual references. Specification of details is not necessary here. Let it suffice to explain that **S** expresses the successor function which assigns to every natural number n the next number n+1. **S** interprets all symbols and symbols only. It is convenient to combine readability of usual notation of expressions like **0≠1** *(zero differs from one)* with the simplicity of a reduced vocabulary like in ¬**0** = **S0** *(it is not the case that zero equals the successor of zero)* and Polish syntax like in ¬= **0S0** by freely switching between them. Tarski's so-called definition of arithmetical satisfiability extends **S** to complex terms and formulas. For the sake of clarity his extension may be symbolized by **S*** and defined in the well known way which need not be repeated. In fact, it would be better not to look upon **S*** as defining satisfaction but as the crucial step towards a definition. The main point of the following argument emerges from the possibility of introducing an intermediate kind of interpretation **I** in the style of (9):

9.

> *Let A be an n-place function symbol, predicate symbol, or connective sign and $B_1, \ldots, B_n$ the corresponding argument expressions.*
> $\mathbf{I}[A\,B_1 \ldots B_n] = \langle \mathbf{S}[A], \mathbf{I}[B_1], \ldots, \mathbf{I}[B_n] \rangle.$

Here names count as 0-place function symbols, propositional symbols as 0-place predicate symbols, 0-place truth connectives do not occur. Of course, ordinary function symbols and predicate symbols take terms as argument expressions, connective signs take formulas. Predicate symbols are not interpreted by sets of n-tuples, but in Frege's manner as characteristic functions. In order to retain simplicity the argument expressions are not included in brackets. The novel assignment **I** as well as the classical Tarskian interpretation **S*** extends **S** from symbols to complexes. In contrast to **S*** the **I**-assignment does not yield an object or a truth value or better satisfaction value, but applied to a sequence of symbols it simply yields the sequence of interpretations of the symbols according to **S**, where the functions are not applied to arguments to generate values. The **I**-interpretation of a sequence of symbols is the congruent sequence of **S**-interpretations of the symbols and this is what Frege's theory of functions refused to recognize. In the following demonstration ν is the negation function (where $\nu(1){=}0$ and $\nu(0){=}1$), $\chi_=$ is the characteristic function of identity and 0 is the number zero. (10) is the proof that the **I**-interpretation of **0 ≠ 0**, i.e. ¬**0** = **0** or in Polish notation ¬= **00**, is the sequence $\langle \nu, \chi_=, 0, 0 \rangle$.

10.

   a) $\mathbf{I}[\neg{=}\,\mathbf{00}] = \qquad \langle \mathbf{S}[\neg], \mathbf{I}[{=}\,\mathbf{00}] \rangle$ 9

b)	$\langle \nu, \mathbf{I}[= \mathbf{00}] \rangle$	S
c)	$\langle \nu, \mathbf{S}[=], \mathbf{I}[0], \mathbf{I}[0] \rangle$	9
d)	$\langle \nu, \chi_=, \mathbf{S}[0], \mathbf{S}[0] \rangle$	S
e)	$\langle \nu, \chi_=, 0, 0 \rangle$	S

The **I**-assignment does part of the job of **S***, wherefore it is natural to relieve the latter and to symbolize the rest by **E**.

11.

Let ϕ be an n-place function and $\alpha_1, \ldots, \alpha_n$ objects.
$$\mathbf{E}[\langle \phi, \alpha_1, \ldots, \alpha_n \rangle] = \phi(\mathbf{E}[\alpha_1], \ldots, \mathbf{E}[\alpha_n])$$

Clause (11) applies functions ϕ to their arguments $\alpha_1, \ldots, \alpha_n$ to generate values. Accordingly computation (10) continues as in (12):

12.

a)	$\mathbf{E}[\langle \nu, \chi_=, 0, 0 \rangle] =$	$\nu(\mathbf{E}[\langle \chi_=, 0, 0 \rangle])$	11
b)		$\nu(\chi_=(\mathbf{E}[0], \mathbf{E}[0]))$	11
c)		$\nu(\chi_=(0, 0))$	11
d)		$\nu(1)$	$\chi_=$
e)		0	ν

In the step from (b) to (c) the application of **E** to the object 0 yields the object 0. This is in harmony with rule (11) which determines $\mathbf{E}[\phi] = \phi$ for 0-place functions ϕ like names. The truth value of $\neg= \mathbf{00}$, $\neg 0 = 0$ or $0 \neq 0$ turns out to be 0, i.e. falsehood.

Clauses (9) and (11) amount to a novel but natural division of Tarski's extended interpretation **S*** to assignments **I** and **E**. Its meaning results from the possibility to take sequences **I**[A] of interpretations congruent with expressions A as models for the senses or thoughts expressed by A and **E**[**I**[A]] as references of the senses **I**[A] (or of expressions A itself). The sense or thought expressed by $\neg 0 = 0$, i.e. the non-identity of zero with itself, according to (9) is the sequence $\langle \nu, \chi_=, 0, 0 \rangle$ and the reference of this sense according to (11) falsehood. Let us be cautious to distinguish Frege's intuitively conceived senses from sequences generated by the **I**-interpretation, and refer to the latter as **I**ntensions and to the values of **E** naturally as Extensions. According to (9) Intensions of single symbols like **f** oder **a** are simple and coincide with their Extensions. Accordingly, it is not wrong to speak of Intensions of symbols but pointless. In case of complex terms or formulas Intensions are likewise complex and this accounts for the widely shared intuition according to which senses necessarily have an internal machinery or structure. In this model Intensions owe complexity to their expressions. Word formations like *evening star* and *morning star* have an unclear status which arises the interest of linguists but not of logicians. We have to decide whether they count as simple or complex. In the former case their Intensions and Extensions coincide, in the latter case expressions communicate their complexity to their

Intensions. This move amounts again to a division of semantics. In contrast to Frege it does not split the I-interpretation of words or symbols into senses and references, but Tarski's I*-interpretation into Intensions and Extensions. One of the results is that Tarski's semantics no longer appears purely extensional and that the senses only remained incognito.

The main point of the difference between **E** and **I** is stated in (13).

13.

> The **S**- and **I**-assignments are trivial, but not the **E**-assignment and **S***.

Let us sketch a proof.[11] The arguments of **S** are symbols which can be numbered in some arbitrary way. The values of **S** are objects and intuitively computable functions.

14.

> $S[0] = 0$, i.e. zero
> $S[S] = s$, i.e. the successor function $(s(n)=n+1)$
> $S[+] = +$, i.e. addition
> $S[\times] = \times$, i.e. multiplication
> $S[=] = \chi_=$, i.e. the characteristic function of identity
> $S[\neg] = \nu$, i.e. the negation function $(\nu(1)=0$ and $\nu(0)=1)$
> $S[\wedge] = \kappa$, i.e. the conjunction function
> $(\kappa(x, y) = 1$ if and only if both x=1 and y=1)

Computable functions are recursively enumerable and can be indexed by natural numbers. **S** can be coded as a finite assignment of numbers to numbers and therefore it is trivial. Because of recursive decidability of syntax, triviality of **S** entails triviality of the **I**-assignment and its representability in arithmetic. In contrast, if **S*** were representable in arithmetic, truth would be definable in arithmetic which is not the case according to Tarski's theorem. Moreover, if **E** were representable **EI** = **S*** would be so. Because all recursive functions are representable and all recursive sets are definable in arithmetic the notion of arithmetical truth is not recursively definable and **E**-assignment is not recursively computable.

Intensions are understood easily and the determination of Extensions meets growing difficulties. In this essential respect this pair of concepts resembles Frege's sense and reference. The advantage of the I-E- distinction compared to Frege's distinction stems from several facts. Firstly, all syntactically different expressions are equipped with different Intensions or in the case of sentences with different I-thoughts or I-propositions independently of their equivalence or lack of equivalence. Only in the special cases of differences resulting exclusively from substitution of synonymous words or symbols are the same Intensions bestowed. Secondly, whereas Frege's senses should account

---

[11] Alfred Tarski 1936 & 1956

for both comprehension and difficulty of terms and sentences, which proved impossible here, Intensions enable comprehension and difficulties arise with respect to Extensions. Thirdly, the theory of recursive functions and complexity theory provides an analysis of the nature of difficulties. But there is not room to expound on this here. Let us close by repeating that the I–E division clearly applies to syntactically complex expressions and that Frege probably had in mind the same distinction, but unfortunately was seduced by a tendency to simplicity and read the distinction into words. He arrived at a division of the semantics S of words and symbols into sense and reference and touched on the unanalyzable.

References

Currie, G. (1982): *Frege. An Introduction To His Philosophy* (Sussex & New Jersey)

Frege, G. (1892): *Über Sinn und Bedeutung*. Zeitschrift für Philosophie und philosophische Kritik, NF 100, p. 25–50 [Angelelli p. 143–62, Patzig p. 40–65, English translation in Geach & Black p. 56–78, Reprinted in Moore, p. 23–42]

Frege, G. (1952): Translations from the Philosophical Writings of Gottlob Frege, ed. Peter Geach and Max Black (Basil, Blackwell &Mott, Oxford)

Frege, G. (1976): *Kleine Schriften*, ed. Ignacio Angelelli (Wissenschaftliche Buchgesellschaft, Darmstadt)

Frege, G. (1976): *Wissenschaftlicher Briefwechsel*, herausgegeben, bearbeitet, eingeleitet und mit Anmerkungen versehen von Gottfried Gabriel, Hans Hermes, Friedrich Kambartel, Christian Thiel und Albert Veraart (Hamburg)

Frege, G. (1986): *Funktion, Begriff, Bedeutung. Fünf logische Studien*, ed. G. Patzig. 6. Aufl. (Vandenhoek & Ruprecht, Göttingen)

Frege, G. (1991): *Funktion und Begriff* (Hermann Pohle, Jena) (Frege 1976, p. 125–142)

Kemmerling, A. (1990): *Gedanken und ihre Teile*, Grazer Philosophische Studien, ed. Rudolf Haller 37, p. 1–30

Moore, A. W. (ed) (1993) *Meaning and Reference* (Oxford University Press, Oxford)

Putnam, H. (1973): *Meaning and Reference*, The Journal of Philosophy 70, p. 699–711

Tarski, A. (1936): *Der Wahrheitsbegriff in den formalisierten Sprachen*, Studia Philosophica 1, p. 261–405, English translation with the title *The Concept of Truth in Formalized Languages* in Tarski (1956) p. 152–278

Tarski, A. (1953): *Undecidable Theories*, in collaboration with Andrzej Mostowski & Raphael M. Robinson (North Holland Publishing Company, Amsterdam)

Tarski, A. (1956): *Logic, Semantics, Metamathematics. Papers from 1923 to 1938*, translated by J.H. Woodger (Clarendon Press, Oxford)

Logics of Knowing and Believing

Petr Hájek

Institute of Computer Science, Academy of Sciences of the C. R.,
Pod vodárenskou vezi 2, 18207 Prague 8, Czech Republic. e-mail: hajek@uivt.cas.cz

1 Introduction

What is the relation of intelligence (natural or artificial) to *logic*? This is surely a fascinating question with many aspects. We shall not discuss what intelligence is; but we shall start by saying that logic is understood as the study of *consequence* and of *deduction* as obtaining consequences from (accepted, assumed) premises (axioms). It goes without saying that intelligence has a logical aspect (among various other aspects). Intelligent systems are said to work with various sorts of *knowledge*; knowledge representation and processing, and even reasoning about knowledge are considered to be important topics in AI. Knowledge may be *uncertain*; thus one has to work with *beliefs*. It has turned out that logical aspects of reasoning about knowledge and beliefs are well formalized by means of some systems of *modal logic* called logic of knowledge and of belief respectively. An excellent survey of propositional (Boolean) logics of knowledge and belief in relation to computational complexity (with full proofs of main results) is given by Halpern (1992). Most basic facts on these logics are surveyed in Sect. 2 of this chapter. Section 3 surveys basic facts on predicate (Boolean) logics of knowledge and belief; Hughess and Cresswell (1984) contains details. The main part is Sect. 4 where we take into consideration the fact that most knowledge in AI systems (not speaking about natural intelligence) is *imprecise* (vague, fuzzy), which leads us to *fuzzy logic*. The reader should not be shocked by this term and should distinguish, from the beginning, between two possible meanings: broad – it is a current fashion to call everything concerning fuzziness just fuzzy logic – and *narrow*: logical calculi of deduction from vague premises. This is the meaning we shall understand. And in has turned out that these calculi are some *many-valued logics*: sentences may have a truth degree which can be any real number between 0 and 1.

It is very important to distinguish carefully between uncertainty as degree of belief (probability, etc.) on the one hand and vagueness (fuzziness) as degree of truth on the other. I refer to Hájek (1995a) for a survey of calculi of fuzzy logic where I also discuss the mentioned difference in more detail. But, as also mentioned there, on can build (and use) *combined* calculi of both uncertainty and vagueness, i.e., many-valued modal logical systems. In Sect. 4 we first generalize modal logics of knowledge and belief to the many-valued case and then obtain some quantitative (and comparative) systems of possibilistic logic as a particular case. Section 5 contains some comments and open problems.

The reader is assumed to have basic familiarity with the classical (Boolean) propositional logic, i.e., to know how formulas are built from propositional variables and connectives (implication and negation, say), how other connectives (conjunction and disjunction, equivalence) are defined from basic connectives, what the truth tables of connectives are and how an evaluation of propositional variables by truth values 1 (truth) and 0 (falsity) extends to the corresponding evaluation of all propositional formulas; a tautology is a formula having the truth value 1 for each evaluation of propositional variables, i.e., being identically true. Some tautologies are taken for logical axioms and the notion of a proof is defined; the completeness theorem says that a formula is a tautology iff it has a proof. For other calculi the structure is similar: we define semantics (what is a tautology) and deductive structure (provability) and ask whether we have completeness, i.e., whether identical truth coincides with provability. If the reader knows how this is elaborated for the case of the classical (Boolean) predicate calculus (i.e., logic with predicates and quantifiers), he/she can enjoy Sect. 3; otherwise this section may be skipped.

2   Propositional Logics of Knowledge and Belief

The language of a modal propositional logics consists of propositional variables $p, q, \ldots$, connectives (say implication $\rightarrow$ and negation $\neg$) and the modality $\Box$. *Formulas* are defined as follows:

(i) each propositional variable is a formula;
(ii) if φ, ψ are formulas then $\neg\varphi$, $\varphi \rightarrow \psi$, $\Box\varphi$ are formulas;
(iii) each formula results from propositional variables by iterated use of (ii).

Thus e.g. $\Box q \rightarrow \Box(p \rightarrow q)$ is a formula. For each φ, the formula $\Box\varphi$ is read "necessarily φ"; in particular systems we may prefer other readings, namely "it is known that φ" or "it is believed that φ". We define other connectives as usual, i.e., $\varphi \wedge \psi$ stands for $\neg(\varphi \rightarrow \neg\psi)$, $\varphi \vee \psi$ stands for $\neg\varphi \rightarrow \psi$, $\varphi \equiv \psi$ stands for $(\varphi \rightarrow \psi) \wedge (\psi \rightarrow \varphi)$. We define the modality $\Diamond$ of possibility; $\Diamond\varphi$ stands for $\neg\Box\neg\varphi$ (φ is possible if $\neg\varphi$ is not necessary). If we read $\Box$ as "it is known (believed) that ..." then $\Diamond\varphi$ may be read "φ is admitted".

The semantics is based on the idea of a set W of *possible worlds*; the truth value of a propositional variable p depends on possible worlds: p may be true in some worlds, false in some other. More precisely, we define a *Kripke model*[1] of the logic of knowledge to be a pair $\langle W, e \rangle$ where W is a non-empty set (of possible worlds) and e is a mapping assigning to each propositional variable p and each world $w \in W$ a truth value $e(w, p) = 0$ or 1. The truth value $\|\varphi\|_w$ of a formula φ in a world w is defined as follows:

---

[1] See Kripke (1954), Kripke (1962), Kripke (1963) for Kripke's pioneering work.

if φ is a variable p then $\|\varphi\|_w = e(w,p)$;
$\|\neg\psi\|_w = 1$ iff $\|\psi\|_w = 0$;
$\|\psi \to \chi\|_w = 1$ iff $\|\psi\|_w = 0$ or $\|\chi\|_w = 1$;
$\|\Box\psi\|_w = 1$ iff $\|\psi\|_v = 1$ for all $v \in W$.

Thus for connectives we use "truth tables"; $\Box\psi$ is true (in a world w) if ψ is true in *all* possible worlds. (Hence $\Box\psi$ is either true in all worlds or in none.) This corresponds well to the idea of *knowing* ψ: ψ is *known* to be true if it is true in all possible worlds. The idea is that one of possible worlds is the actual world, but we may not know which one. Nevertheless if ψ is true in *all* possible worlds (of the given Kripke model) then ψ is *known* to be true. For historical reasons, the logic of knowledge is denoted S5.

Example. Let $W = 1, 2, 3$, take two propositional variables and let e be given by the following table:

	p	q
1	1	1
2	1	0
3	0	1

Observe that the formula $\Box(\varphi \vee \psi)$ is true (in any world of this model) but $\Box\varphi$, $\Box, \psi$ are false; thus $\Diamond\neg\varphi$, $\Diamond\neg\psi$ are true.

A formula φ is a *tautology* of S5 if it is true in each world of each Kripke model of S5. The following are examples of tautologies:

(1) $\Box(\varphi \to \psi) \to (\Box\varphi \to \Box\psi)$,
(2) $\Box\varphi \to \Box\Box\varphi$,
(3) $\Diamond\varphi \to \Box\Diamond\varphi$, (i.e., $\neg\Box\neg\varphi \to \Box\neg\Box\neg\varphi$),
(4) $\Box\varphi \to \varphi$.

Let us read them speaking about knowledge: (1) says that if you know that φ implies ψ then if you (also) know that φ then you know that ψ. (2) is called "positive introspection": if you know that φ then you know that you know that φ. (3) is negative introspection: if you do not know that $\neg\varphi$ then you know that you do not know that $\neg\varphi$. (in other words: if you admit φ then you know that you admit φ). (4) says that if you know that φ then φ (is true).

These are indeed plausible properties of knowing. More than that: take (1)–(4) for *axioms* of S5, together with some axioms of propositional calculus, e.g. the following:

(01) $\varphi \to (\psi \to \varphi)$
(02) $\varphi \to (\psi \to \chi)) \to ((\varphi \to \psi) \to (\varphi \to \chi))$
(03) $(\neg\varphi \to \neg\psi) \to (\psi \to \varphi)$.

Define the *deduction rules*:
– modus ponens: from φ and $\varphi \to \psi$ deduce ψ;
– necessitation: from φ deduce $\Box\varphi$.

Define a *proof* (in S5) to be a sequence $\varphi_1, \ldots, \varphi_n$ of formulas whose each member φ_i either is an axiom (of one the forms (1)–(4), (01)–(03)) or is deduced from some preceding members of the sequence by one of the deduction rules.

Define a formula φ to be *provable* (in S5) if it is the last member of a proof (in S5).

It is easy to show that this deductive system is *sound*, i.e., each (S5)-provable formula is a tautology (of S5). The converse is the *completeness theorem*: if φ is a tautology of S5 then φ is S5-provable.

*

Now we turn to the *logic of belief*. I believe that φ if φ is true in all worlds I consider believable; but it may happen that the "actual" world is in fact not among them so that it can happen that φ is false in it. This leads to the following variant of modal logic (we shall call it the logic of belief and denote, for historical reasons, KD45):

The formulas are as in S5. A *Kripke model* of KD45 has the form $K = \langle W, e, W_0 \rangle$ where W is an non-empty set of possible worlds, e is an evaluation as above and W_0 is a non-empty subset of W; elements of W_0 are believable worlds. The definition of $\|\Box\psi\|_w = 1$ if $\|\varphi\|_{w'} = 1$ for all $w' \in W_0$. It is immediately seen that again $\|\Box\psi\|_w = 1$ for all $w \in W$ or for no $w \in W$; but we also see that $\|\Box\psi\|_w$ may be 1 whereas $\|\psi\|_w = 0$ if $w \in W-W_0$. In the example above, let $W_0 = \{1, 2\}$; then $\|\Box p\|_3 = 1$ but $\|p\|_3 = 0$.

The axioms of KD45 result from those of S5 by replacing the axiom (4) by

(4′) $\Box\varphi \to \Diamond\varphi.$

All axioms are tautologies of KD45, thus the logic is sound. Moreover, it is *complete*: each tautology of KD45 is provable in KD45.

Just to show an example of a proof in KD45 let us indicate that the formula $\Box(\Box\varphi \to \varphi)$ is provable in KD45. (We only mention main steps since we do not intend to elaborate KD45 systematically.) First, we can prove $(\neg\Box\varphi) \vee (\Box\varphi)$ (this is *tertium non datur*), thus $\Diamond\neg\varphi \vee \Box\varphi$. This gives $\Box\Diamond\neg\varphi \vee \Box\varphi$ by negative introspection. From this we get $\Box(\Diamond\neg\varphi \vee \varphi)$ (since KD45 proves $\Box p \to \Box(p \vee q)$, $\Box q \to \Box(p \vee q)$, thus it proves $(\Box p \vee \Box q) \to \Box(p \vee q)$). But the last formula is equivalent to $\Box(\Box\varphi \to \varphi)$ (since $\neg p \vee q$ is equivalent to $p \to q$).

From this we get a nice *theorem* on the relation between S5, KD45: For each φ, S5 proves φ iff KD45 proves $\Box\varphi$.

Now we show that KD45 has a faithful interpretation in S5. Let b be a new propositional variable. For formulas φ not containing b let φ^* be defined as follows:

p^* is p, $(\psi \to \chi)^*$ is $\psi^* \to \chi^*$, $(\neg\psi)^*$ is $\neg(\psi^*)$,
$(\Box\psi)^*$ is $\Box(b \to \psi)$.

Furthermore, let φ^{**} be $\Diamond b \rightarrow \varphi^*$. Then we get the following:

Theorem. For each formula φ not containing the new propositional variable b, KD45 proves φ iff S5 proves φ^{**}.

The proof is easy; the main idea is that b just defines the set W_0 – the subset of believable worlds – in a model $K = \langle W, e \rangle$ of S5 ($W_0 = \{x \in W \mid \|b\|_x = 1\}$). If $\Diamond b$ is true in W then the set W_0 is non-empty, so that $\langle W, e, W_0 \rangle$ is a model for KD45.

We close this section by two technical remarks (that may be omitted by a reader who is in haste).

(1) Our notion of a Kripke model is a particular case of a more general definition, namely: a Kripke model is $K = \langle W, e, R \rangle$, where K and e are as above and $R \subseteq W \times W$ is a binary relation (of alternativeness). $\|\Box \psi\|_w = 1$ iff $\|\psi\|_v = 1$ for all v such that $R(w, v)$ ($\Box \psi$ is true in w it ψ is true in all worlds alternative to w). Our models of S5 are then structures of the form $\langle W, e, W \times W \rangle$ (all worlds are mutually alternative) and models of KD45 have the form $\langle W, e, W \times W_0 \rangle$.

(2) Properties of the relation R are important for truth of important formulas in the Kripke model. Note that if $K = \langle W, e, R \rangle$ is a Kripke model in which R is an equivalence (reflexive, symmetric and transitive relation) then satisfies all axioms of S5; if R is transitive, serial (each world has at least one alternative) and Euclidean (if u, v are alternatives of w then they are alternatives of each other) then K satisfies all axioms of KD45. But we shall be happy with the models described above.

3   Predicate (Boolean) Logics of Knowledge and Belief

The reader of this section is assumed to know the basics of classical (Boolean) predicate calculus. There are several ways of developing modal predicate logic; we offer just one elegant approach. (More can be found in several articles in Gabbay and Guenther (1984).)

The *language* consists of predicates $P, Q, \ldots$ (each being assigned a positive natural number – its *arity*), object variables $x, y, \ldots$, connectives $\rightarrow, \neg$, the universal quantifier $\forall$, and the modality $\Box$. *Atomic formulas* have the form $P(x_1, \ldots, x_n)$ where P is an n-ary predicate and $x_1, \ldots, x_n$ are object variables. Other formulas are built according to the rules

– if φ, ψ are formulas then $\varphi \rightarrow \psi$, $\neg \varphi$, $\Box, \varphi$ are formulas;
– if φ is a formula and x an object variable then $(\forall x)\varphi$ is a formula.

Connectives $\wedge, \vee, \equiv$ and the modality $\Diamond$ are introduced as above; the existential quantifier $\exists$ is introduced by declaring $(\exists x)\varphi$ to be an abbreviation for $\neg(\exists x)\neg\varphi$.

A *Kripke structure* for predicate S5 has the form $K = \langle W, M, r_P, r_Q, \ldots \rangle$ where W is a non-empty set of objects and for each predicate P of arity n, r_P is a subset of $M^n \times W$ (M^n is $M \times \ldots \times M$, n factors). The fact that an

$(n+1)$-tuple $\langle a_1, \ldots, a_n, w \rangle$ is in r_P means that in the world w, $\langle a_1, \ldots, a_n \rangle$ is in the relation interpreting P. (We could say that r_P is a system of n-ary relations on M, indexed by elements of w.)

The specific assumption we make here is that the set of objects is fixed, independently of possible worlds. This admits generalizations, but we shall disregard them.

Given a Kripke model $K = \langle W, M, (r_P)_P \rangle$ and an evaluation S of object variables by elements of M, we define the *value* $\|\varphi\|_w^K[S]$ of a formula φ in a world w under the evaluation S. Our definition is a standard Tarski-style definition, by induction on the complexity of formulas.

$$\|P(x_1, \ldots, x_n)\|_S^K[w] = 1 \text{ iff } \langle S(x_1), \ldots, S(x_n), w \rangle \in r_P$$

($P(x_1, \ldots, x_n)$ is satisfied if the tuple of values of $x_1, \ldots, x_n$ is in the relation interpreting P in the world w);

$$\|\neg\varphi\|_S^K[w] = 1 \text{ iff } \|\varphi\|_S^K[w] = 0,$$
$$\|\varphi \to \psi\|_S^K[w] = 1 \text{ iff } \|\varphi\|_S^K[w] = 0 \text{ or } \|\psi\|_S^K[w] = 1,$$
$$\|\Box\varphi\|_S^K[w] = 1 \text{ iff } \|\varphi\|_S^K[v] = 1 \text{ for all } v \in W,$$
$$\|(\forall x)\varphi\|_S^K[w] = 1 \text{ iff } \|\varphi\|_{S'}^K[w] = 1 \text{ for all evaluations } S' \text{ differing from } S$$
$$\text{at the most in the value of } x.$$

A formula φ is *true* in K if $\|\varphi\|_S^K[w] = 1$ for all S, w. Furthermore, φ is a *tautology* (of predicate Boolean S5) if φ is true in each Kripke model (of predicate Boolean S5).

One can easily verify that all axioms of propositional S5 (i.e., (1)–(4), (01)–(03)) are tautologies of predicate Boolean S5; moreover, the usual axioms for quantifiers in predicate logic, i.e.,

(04) $(\forall x)\varphi(x) \to \varphi(t)$ (if t is free for x in φ)
(05) $(\forall x)(\chi \to \varphi) \to (\chi \to (\forall x)\varphi)$ (if x is not free in χ)

are tautologies of our logic. Modus ponens and necessitation preserve tautologicity; and so does *generalization* (from φ deduce $(\forall x)\varphi$). Thus, with the obvious notion of proof and provability (axioms (01)–(05), (1)–(4), deduction rules modus ponens, necessitation and generalization) this logic is sound: each provable formula is a tautology. Moreover, we have completeness:

A formula φ is provable in predicate Boolean S5 iff it is a tautology.

In particular, the following formula, called the *Barcan formula* is provable in our logic:

(B) $$\Box(\forall x)\varphi = (\forall x)\Box\varphi.$$

Thus $\forall$ commutes with $\Box$.

*

For predicate Boolean KD45 we change the notion of a Kripke model by adding a non-empty set $W_0 \subseteq W$ of *believable worlds*:

$K = \langle W, M, (r_P)_{P\,\text{predicate}}, W_0 \rangle$ and change the definition of satisfaction as follows:

$$\|\Box\psi\|_w = 1 \text{ iff } \|\psi\|_v = 1 \text{ for all } v \in W_0.$$

The corresponding axiom system of predicate Boolean KD45 contains axioms of predicate calculus (01)–(05), axioms (1), (2), (3), (4') of KD45 *plus* the Barcan formula. One can show that this system is *sound and complete* with respect to tautologies of predicate Boolean KD45, i.e., formulas true in each Kripke model $K = \langle W, M, (r_P)_P, W_0 \rangle$ in the sense of our modified definition of satisfaction.

This completes our (telegraphic) survey of predicate (Boolean, i.e., two-valued) modal logic of knowledge and belief.

4   Fuzzy Propositional Logics of Knowledge and Belief

A fuzzy propositional logic differs from the Boolean propositional logic by having more than two truth values, 0 and 1. Here 0 stands for (absolute) falsity and 1 for (absolute) truth; but there are intermediate truth degrees, finitely or infinitely many. (They are assumed to be linearly ordered, with 0 as the least and 1 as the largest element.) An evaluation e of propositional variables assigns to each variable a truth value; one has *connectives* ($\rightarrow, \neg$ and possibly others) and has to define their *truth functions*, allowing one to compute the truth value of a compound formula from the truth values of its components. There are some natural conditions on truth functions: first if all, a truth function of implication (conjunction, etc.) must give for assignments taking only values zero and/or one classical (Boolean) values of implication (conjunction etc.). The second natural condition is monotonicity: e.g. the truth function of implication should be non-increasing in the first argument and non-decreasing in the second. For modal logics over finitely valued propositional calculi see Fitting (1992a), Fitting (1992b); we shall concentrate on logics whose set of truth values is the real interval $[0,1]$. See Hájek (1995a) for a survey; it turns out that there are three basic real-valued propositional logics, each given by a truth-function for conjunction. Here we rely on Pavelka-style extension of Łukasiewicz's propositional logic (see also Hájek (1995b)). We shall directly describe the fuzzy variant of S5. The next part of this section is a shortened version of Hájek and Harmancová (without proofs).

*

First we give a detailed description of our logic.

The *truth values* are reals from the unit interval [0,1].

The *language* consists of propositional variables, logical connectives $\rightarrow, \neg$ (other connectives $\&, \underline{\vee}, \wedge, \vee$ are defined), truth constant $\bar{r}$ for each *rational* r, modality $\Box$ ($\Diamond$ definable). The definitions are as follows:

$\varphi \& \psi$ stands for $\neg(\varphi \to \neg \psi)$,

$\varphi \underline{\vee} \psi$ stands for $\neg \varphi \to \psi$,

$\varphi \vee \psi$ stands for $(\varphi \to \psi) \to \psi$,

$\varphi \wedge \psi$ stands for $\neg(\varphi \to \neg(\varphi \to \psi))$,

$\Diamond \varphi$ stands for $\neg \Box \neg \varphi$.

The *truth functions* of connectives (denoted by the same symbol as the connective but with a bullet) are: $x \to^{\bullet} y = \min(1, 1 - x + y)$, $\neg^{\bullet} x = 1 - x$, $x \&^{\bullet} y = \max(0, x + y - 1)$, $x \underline{\vee}^{\bullet} y = \min(1, x + y)$, $x \vee^{\bullet} y = \max(x, y), x \wedge^{\bullet} y = \min(x, y)$.

Formulas are defined as usual; *semantics* is given by Kripke models of the form $K = \langle W, e \rangle$ where $W \neq \emptyset$ is a set of possible worlds and $e : (Var \times W) \to [0, 1]$ is an evaluation assigning to each propositional variable p and each world $w \in W$ a truth value $e(p, w)$. (Var is the set of all propositional variables.) The evaluation e extends to all formulas as usual:

$\|\bar{r}\|_{K,w} = r$,

$\|\varphi\|_{K,w} = e(\varphi, w)$ for φ atomic,

$\|\varphi \to \psi\|_{K,w} = \|\varphi\|_{K,w} \to^{\bullet} \|\psi\|_{K,w}$,

$\|\neg \varphi\|_{K,w} = \neg^{\bullet} \|\varphi\|_{K,w}$,

$\|\Box \varphi\|_{K,w} = \|\Box \varphi\|_{K} = \inf_{w' \in W} \|\varphi\|_{K,w'}$.

A formula φ is a *1-tautology* if $\|\varphi\|_{K,w} = 1$ for each Kripke model $K = \langle W, e \rangle$ and each $w \in W$.

Logical axioms:

Axioms of Łukasiewicz's propositional logic:

$\varphi \to (\psi \to \varphi)$,

$(\varphi \to \psi) \to ((\psi \to \chi) \to (\varphi \to \chi))$,

$(\neg \varphi \to \neg \psi) \to (\psi \to \varphi)$,

$((\varphi \to \psi) \to \psi) \to ((\psi \to \varphi) \to \varphi)$,

Axioms of S5:

$\Box(\varphi \to \psi) \to (\Box \varphi \to \Box \psi)$,

$\Box \varphi \to \Box \Box \varphi$,

$\Diamond \varphi \to \Box \Diamond \varphi$,

$\Box \varphi \to \varphi$,

Bookkeeping axioms:

$(\bar{r} \to \bar{s}) \equiv \overline{r \to^{\bullet} s}$,

$\neg \bar{r} \equiv \overline{\neg^{\bullet} r}$,

$\bar{r} \equiv \Box \bar{r}$,

Fitting-style axioms:

$(\bar{r} \to \Box \varphi) \equiv \Box(\bar{r} \to \varphi)$,

$(\bar{r} \to \Diamond \varphi) \equiv \Diamond(\bar{r} \to \varphi)$.

Last axiom:

$$(\Diamond \varphi \mathbin{\&} \Diamond \varphi) \equiv \Diamond(\varphi \mathbin{\&} \varphi).$$

Deduction rules: Modus ponens and generalization: from φ infer $\Box\varphi$.

A *proof* is a sequence of formulas $\varphi_1, ..., \varphi_n$ such that for each $j = 1, ...n$, φ_j is an axiom or follows from some previous formulas by one of the deduction rules. $\vdash \varphi$ means that φ is provable (has a proof).

This completes the definition of the logic; it may be called *fuzzy S5*. Recall that Łukasiewicz's axioms are complete for Łukasiewicz's propositional logic (see e.g. Gottwald (1988) for a detailed proof as well as proofs of many formulas in Łukasiewicz's propositional logic); for a completeness of a simplified Pavelka-style extension of this logic by truth constants see Hájek (1995b) and Hájek (1995a).

Soundness: If φ is provable then φ is a 1-tautology.

We make the following definitions: The *provability degree* of φ is $|\varphi| = \sup\{r \mid \vdash (\bar{r} \to \varphi)\}$. The *truth degree* of φ is $\| \varphi \| = \inf\{\| \varphi \|_{K,w} \mid K = \langle W, e \rangle$ Kripke model and $w \in W\}$.

Strong soundness: $|\varphi| \leq \| \varphi \|$, i.e., whenever $\vdash \bar{r} \to \varphi$ then $\| \varphi \|_{K,w} \geq r$ for each K, w.

Completeness Theorem: For each φ, $|\varphi| = \| \varphi \|$. In particular, φ is a 1-tautology iff, for each $r < 1$, $\vdash \bar{r} \to \varphi$.

Now let us turn to KD45. The natural fuzzification (as investigated in Hájek, Harmancová, Esteva, Garcia, and Godo(1994), Hájek, Harmancová and Verbrugge(1994) for finite-valued logic) is the structure $K = \langle W, \pi, e \rangle$ where π is a *fuzzy* subset of W such that $\sup_w \pi(w) = 1$, (and e is a $[0, 1]$-valued evaluation of atoms); we put $\| \Diamond\varphi \|_{K,w} = \sup_{w'}(\pi(w') \wedge^\bullet \| \varphi \|_{K,w'})$, thus $\| \Box\varphi \|_{K,w} = \inf_{w'}(\neg^\bullet\pi(w') \vee^\bullet \| \varphi \|_{K,w'})$. Then axioms of KD45 are tautologies (see the corresponding proofs in Hájek, Harmancová, Esteva, Garcia, and Godo(1994)).

Each model $K = \langle W, \pi, e \rangle$ of fuzzy KD45 determines a model $K' = \langle W, e' \rangle$ of fuzzy S5 with the language extended by a new propositional variable q such that e' extends e and $e'(q, w) = \pi(w)$. The modalities of MKVD45 become $\Diamond_q\varphi = \Diamond(q \wedge \varphi)$, $\Box_q\varphi = \Box(\neg q \vee \varphi)$; note $\| \Diamond q \|_{K'} = 1$.

Clearly, the above construction of K' from K shows that MKVD5 has a faithful interpretation in fuzzy S5 extended by the additional axiom $\Diamond q$. ($\Box, \Diamond$ are interpreted by $\Box_q, \Diamond_q$), i.e., we can understand each formula of fuzzy KD45 as a certain formula of fuzzy S5 (containing and additional propositional variable) such that the formula is a 1-tautology of fuzzy KD45 iff its interpretation is provable in the mentioned strengthening of fuzzy S5. Needless to say, this is only a preliminary result.

*

Let us now consider particular models $K = \langle W, e, \pi \rangle$ of fuzzy KD45 with the evaluation e taking only values 0,1. (But π remains a normalized fuzzy subset of W.) Thus each formula is in each world either true or false; but the formula $\Diamond \varphi$ may have any value in $[0, 1]$. (Recall $\|\Diamond\varphi\|_K = \sup_w \{\pi(w) \wedge^\bullet \|\varphi\|_{K,w}\}$.) The number $\|\Diamond\varphi\|$ is called the (quantitative) *possibility* of φ (given by K) and denoted $\Pi(\varphi)$ (or $\Pi_K(\varphi)$). The function Π satisfies the following: $\Pi(\varphi) = 1$ and $\Pi(\neg\varphi) = 0$ for each tautology φ (of Boolean propositional logic), $\Pi(\varphi \vee \psi) = \max(\Pi(\varphi), \Pi(\psi))$; if φ, ψ are logically equivalent (i.e., $\varphi \equiv \psi$ is a tautology) then $\Pi(\varphi) = \Pi(\psi)$. Conversely, each function Π with there properties is given by an appropriate Kripke model $\langle W, e, \pi \rangle$ with 0-1-valued e. Note that $\Pi(\varphi \wedge \psi) \leq \min(\Pi(\varphi), \Pi(\psi))$ but $\leq$ cannot be replaced by $=$ in general. *Possibility theory* is a well developed theory of uncertainty (see, e.g., Dubois and Prade (1992)); it is a quantitative counterpart of the modality $\Diamond$ and, as we have seen, it embeds into the fuzzy model S5.

One may introduce a binary modality $\lhd$ meaningful in Kripke models $\langle K, e, \pi \rangle$ with two-valued e; $\varphi \lhd \psi$ is true if $\Pi(\varphi) \leq \Pi(\psi)$. This can be nicely axiomatized; see Bendová and Hájek (1993) for details.

Needless to say, one can generalize this for arbitrary Kripke models of the form $\langle K, e, \pi \rangle$ where π is a normalized fuzzy subset of W; see Hájek, Harmancová, Esteva, Garcia, and Godo(1994) for details in the finite valued case.

5 Appendix: Logics of Knowledge and Belief; Their Relation to AI

For the reader having doubts on the relevance of what we elaborate for AI we add some discussion on this question. We content ourselves, more or less, with simply presenting a choice of quotations from some recognized authorities in the domain of AI and related domains.

(1) *Knowledge processing is vital for AI.* Lenat and Feigenbaum (1991): "The Knowledge principle: If a program is to perform a complex task well, it must know a great deal about the world in which it operates. In the absence of knowledge, all you have left is search and reasoning, and this is not enough."

(2) *Logic is vital for knowledge representation.* Nilsson (1991): "The most versatile intelligent machines will represent much of their knowledge about their environments declaratively ... . For the most versatile machines, the language in which declarative knowledge is represented must be at least as expressive as the first-order predicate calculus."

Winograd and Flores (1986): "Research in AI has emphasized the problem of representation. ... In general, AI researchers make use of formal logical systems (such as predicate calculus) for which the available operations and their consequences are well understood. ... There are operations that manipulate the symbols in such a way as to produce veridical results." (p. 85)

(3) *Intelligent systems should be able to reason about their own knowledge.*
Winograd and Flores (1986): "Whenever we treat a situation as present-at-hand, analyzing it in terms of objects and their properties, we thereby create a blindness. Our view is limited to what can be expressed in terms we have adopted." (p. 97) Let us note that Winograd and Flores do not say this in connection with logics of knowledge and belief; their criticism is much more fundamental (and concerns what they call the rationalist tradition). But the fact that simple-minded AI systems are unable to recognize their own limits has been stressed for a long time. Hayes-Roth et al. (1983): "Today's expert systems ... are unable to recognize or deal with problems for which their own knowledge is inapplicable or insufficient." (p. 55) Formalizing logics of knowledge is one of possible ways to fight against the blindness created. (No final victory through logic of knowledge is expected; the new system creates its own blindness. But at least this is one step forward.) Halpern (1992): "Reasoning about knowledge and belief has long been an issue of concern in philosophy and artificial intelligence. ... In order to formally reason about knowledge, we need a good semantic model. Part of the difficulty in providing such a model is that there is no agreement on exactly what properties of knowledge are or should be. Do you know what you don't know? Do you know only true things or can something you 'know' actually be false?"

(4) *The knowledge of AI systems is typically uncertain and imprecise.* Hayes-Roth et al. (1983): "All of the data may not be available, some may be suspect, and some of the knowledge for interpreting the data may be unreliable. These difficulties are part of the normal state of affairs in many interpretation and uncertain and incomplete data has invited a variety of technical approaches." Undoubtedly, logics of belief are one such approach. The presented modal logical approach to belief is surely not the only one possible. The probabilistic approach, or more generally, the approach of numerical beliefs is another way. One of our results is the observation that a fuzzy belief logic naturally gives rise to numerical possibility measures.

(5) *Logics of beliefs as well as probabilistic methods are well established in AI*; for fuzzy logic this is not yet the case – unjustly. To demonstrate this one can inspect the series of proceedings of the well known American annual conferences "Uncertainty in AI". The author shares the opinion that time is ripe for a broad use of fuzzy logic (both in the narrow and the wide sense) in AI. Zadeh, speaking on the growing use of soft computing and especially fuzzy logic in the conception and design of intelligent systems, Zadeh (1993) stated: "The principal aim of soft computing is to exploit the tolerance for imprecision and uncertainty to achieve tractability, robustness, and low production cost. At this juncture, the principal constituents of soft computing (SC) are fuzzy logic (FL), neural networks (NN), and probabilistic reasoning (PR) ... In large measure, FL, NN, and PR are complementary rather than competitive." For a survey of the present state of knowledge on fuzzy logic in the narrow sense see my paper Hájek (1995a). The present contribution

shows, among other things, how a fuzzy logic of knowledge and belief can be constructed in reasonable analogy to the two-valued (Boolean) logic of knowledge and belief. It is hoped that fuzzy logic (in the narrow sense) soon finds its place among the logics of AI.

6 Conclusions and Open Problems

We have surveyed the basic facts about the logic S5 (of knowledge) and KD45 (of belief); we also stated their versions over the predicate calculus and over the fuzzy propositional calculus. Admittedly, these logics do formalize some important properties of knowing and believing; and the fuzzified version leads to the (well known) numerical theory of possibility. This is a piece of evidence that the systems studied are natural and that creation of a unified logical theory of uncertainty and impreciseness (in the frame of many-valued modal logic) is possible.

Clearly there remain open technical problems, such as

(1) Elegant axiomatization of fuzzy KD45;
(2) Possibility of improvement of the completeness result for fuzzy S5: is each 1-tautology provable? (We have only claimed that if φ is a 1-tautology than $| \varphi | = 1$, i.e., for each $r < 1$, $(\bar{r} \to \varphi)$ is provable.)
(3) Investigation of fuzzy predicate S5, KD45.

These (and other) problems remain a future research task. The papers Fagin (1994), Fine (1985), Voorbraak (1993) and van der Hoek (1992) are recommended, together with those cited above, as further reading.

Acknowledgment. The support of the Forschungsstätte Evangelische Studiengemeinschaft Heidelberg covering the costs of my participation in meetings of the working group "Natürliche und künstliche Intelligenz" is acknowledged. This work was partially supported by the grant No. A1030601/1996 of the Grant Agency of the Academy of Science of the Czech Republic.

References

Bendová, K. and Hájek, P. (1993): Possibilistic logic as tense logic, in: N. Pier et al., ed.: *Qualitative Reasoning and Decision Technologies*, CIMNE, 441–450 (Barcelona)

Dubois, D. and Prade, H. (1992): Possibility theory as a basis for preference propagation in automated reasoning, in: *IEEE Int. Conf. on Fuzzy Systems FUZZ-IEEE'92*, 821–832 (San Diego)

Fagin, R. (1994): A quantitative analysis of modal logic. *Journal of Symbolic Logic* 59: 209–252

Fine, K. (1985): Logics containing K4. *Journal of Symbolic Logic* 50: 619–651

Fitting, M. (1992a): Many-valued modal logics, *Fundamenta Informaticae* 15: 235–254

Fitting, M. (1992b): Many-valued modal logics II, *Fundamenta Informaticae* 17: 55–73

Gabbay, D.M. and Guenther, F., eds. (1984): Handbook of Philosophical Logic vol. II (Reidel, Dordrecht)

Gottwald, S. (1988): *Mehrwertige Logik* (Akademie-Verlag, Berlin)

Hájek, P. (1995): Fuzzy logic from the logical point of view, in: M. Bartošek, J. Staudek, and J. Wiedermann, eds, *SOFSEM'95: Theory and Practice of Informatics; Lecture Notes in Computer Science 1012*, 31–49 (Springer, Berlin-Heidelberg)

Hájek, P. (1995): Fuzzy logic and arithmetical hierarchy, *Fuzzy Sets and Systems* 73(3): 359–363

Hájek, P. (1997): Fuzzy logic and arithmetical hierarchy II, *Studia Logica* 58: 129–141

Hájek, P., Harmancová, D., Esteva, F., Garcia, P. and Godo, L. (1994): On modal logics for qualitative possibility in a fuzzy setting, in: R. López de Mántaras and D. Poole, eds., *Uncertainty in Artificial Intelligence; Proceedings of the Tenth Conference*, 278–285 (Seattle, WA)

Hájek, P., Harmancová, D., Verbrugge, R. (1994): A qualitative fuzzy possibilistic logic, *International Journal of Approximate Reasoning* 12: 1–19

Hájek, P. and Harmancová, D.: A many valued modal logic, in: Proc. IPMU'96, (Granada, Spain), 1021–1024

Halpern, J.Y. (1992): A guide to completeness and complexity for modal logics of knowledge and belief, *Artifical Intelligence* 54: 319–379

Hughess, G.E. and Cresswell, M.J. (1984): *A companion to modal logic* (Methuen, London)

Hayes-Roth, F. et al., eds. (1983): Building expert systems (Addison-Wesley, Reading)

van der Hoek, W. (1992): Modalities for reasoning about knowledge and belief, Thesis, Univ. of Utrecht

Kripke, S. (1954): A completeness theorem in modal logic, *Journal of Symbolic Logic* 24: 1–14

Kripke, S. (1962): Semantic analysis of modal logic I. *Zeitschrift f. Math. Logik und Grundl. der Math.* 9: 67–96

Kripke, S. (1963): Semantic considerations of modal logic, *Acta Phil. Fennica* 16: 83–94

Lenat, D.B. and Feigenbaum, E.A. (1991): On the thresholds of knowledge, *Artificial intelligence* 47: 185–250

Nilsson, N.J. (1991): Logic and Artificial Intelligence, *Artificial Intelligence* 47: 31–56

Voorbraak, F. (1993): As far as I know, Thesis, Univ. of Utrecht

Winograd, T. and Flores, F. (1986): Understanding computers and cognition (Ablex, Norwood)

Zadeh, L.A. (1993): The role of fuzzy logic and soft computing in the conception and design of intelligent systems, in: (Klement and Slany, ed.) Fuzzy logic in AI, Lecture Notes in AI, 695 (Springer, Berlin, Heidelberg) p. 1

Knowledge-Based Systems

Michael M. Richter

FB Informatik, Universität Kaiserslautern, D-67653 Kaiserslautern, Germany.
e-mail: richter@informatik.uni-kl.de

1 Introduction

Efforts to develop knowledge-based systems have now been pursued for over two decades. Languages, inference methods, architectures, and tools have reached a certain standard and have led to interesting and successful applications. Nevertheless, there is still some mystery and confusion among the public about this area. The purpose of this article is to provide a short and simple but nonetheless informative description of the topic including some indications of present trends in research. It should also be remarked that knowledge-based systems (as well as other techniques coming from artificial intelligence) are in no way mysterious and could be understood without knowing what "intelligence" really is. They are nothing more than problem solving methods which have some interesting and (sometimes) new features.

The term knowledge-based system may be somewhat misleading, because it suggests that one is now dealing with expert knowledge and that this was not the case in classical conventional programming. Clearly, classical programs also contained knowledge (sometimes very sophisticated). Also, when a knowledge-based system is translated into machine language there is almost no way to distinguish it from a conventional program. The principle difference in this new development is a different programming style. It is the declarative programming which characterizes AI programming. Declarative programming means that the programmer does not himself have to fix all the details of the execution of the program. Instead he puts down explicitly those facts and laws which in his opinion are necessary and useful for the solution of his problem. Declaritivity also implies that the language in which the domain of interest is described has a fixed semantics in which the meaning of the language constructs is described. The semantics refers to a certain domain of interest called a model; the description of such a model constitutes a large part of the knowledge base.

In order to execute the program the missing details are generated by the control structure and the inference engine. The inference engine deduces logical conclusions from the input knowledge as well as plausible hypotheses and related pieces of information. The control structure organizes this process. The separation of the knowledge base and the inference engine was a major argument in promoting knowledge-based systems: The entries in the knowledge base could vary with the application while the inference engine remains fixed, a fact that should allow a new flexibility compared with classical programming. In this context a knowledge-based system with an empty base was called a shell. Below we will discuss how this view turned out to be very overoptimistic. A knowledge-based system can be considered on three levels which should be clearly separated, the cognitive level, the representation level, and the implementation level.

On the cognitive level the user demands can be expressed. The formal representation on the second level corresponds to a specification in a conventional program. The user demands also describe the functional behavior, i.e., the input-output relation of the system. The functional view is realized and complemented by the architectural view which describes how the functional behavior is realized. The architectural view is realized on the implementation

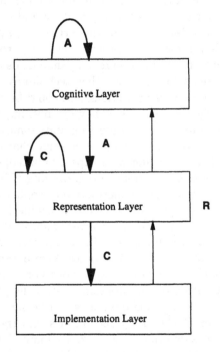

Fig. 1. Between the knowledge utterance and its machine utilization several transformations must be performed (thick arrows). They map in the direction of increased structuring within the layer and proceed from the cognitive form to the formal, and from here to the efficiently processed form. Each syntactic result obtained in the range of a transformation must be associated with its meaning in the domain of the transformation. This is indicated by thin arrows between the layers. The most interesting and difficult one is the inverse mapping into the cognitive layer, which is usually called explanation. The letter A is intended to remind us of "Acquisition" (which is human oriented) while the letter C is short for "Compilation" (which is machine oriented).

and partly on the representation level. In discussing an expert system it is important to point out which functional aspects lead to which architectural problems and how these problems are solved.

An underlying difficulty of the whole knowledge-based approach consists in the transformation between the different layers. Most problems have an immediate and natural representation. This representation, however, is very often not suitable for a good or efficient solution of the problem. The cognitive layer is mainly concerned with such a problem representation. With respect to the solution approach this layer can only reflect how humans would solve the problems. For several reasons this is in general not the way a machine could or should solve the problem: Humans have intuitions and creativity, machines tolerate large quantities of data and complex combinatorial arguments. Therefore a transformation is needed. In the knowledge-based approach this is structured by the acquisition and compilation transformations of Fig. 1.

2   Interpretation and Compilation of Knowledge

For programs in a procedural programming language the semantics is given by a translation to some (abstract) machine on which the program is executed. As mentioned in the introduction, for declarative programs the model theoretic semantics constitutes the meaning. Nevertheless, the declarative programs have to be executed too and therefore they undergo various translation steps indicated in Fig. 1 by the letter C. For such translation one traditionally distinguishes between interpretation and compilation. We will use the term compile time in order to include every action which is performed before an actual problem solving takes place. The three-layer picture then differs from traditional programming only in so far as it gives the human activities a more prominent position. It should be remarked that developments in software engineering presently have the tendency to formally structure the work done by humans in order to establish a software product. From this point of view the difference between the production of knowledge-based systems and classical software systems becomes smaller and smaller. Another change in the methodology of software engineering plays an important role in our context. For many years a piece of software was regarded as the implementation of an algorithm. This static view was responsible for maintenance and related problems: It was difficult and often impossible to react to changes of the specification in order to adapt the algorithm properly. As indicated in the introduction, the separation of knowledge base and inference engine was originally regarded as an important step in overcoming such difficulties. This opinion was based on some hidden assumptions which became clear only later on.

The first assumption was that such a separation can in principle be obtained. More complex problems indicate that such a line of division is rather subjective and hard to motivate; knowledge exists not only about the domain of application but also about ways to obtain a solution, about strategies, etc.

The second major assumption was that the inference methods are in general sufficient to be applied successfully to big classes of problems. Again, experience has shown that this is very often not the case, the shells had either a rather narrow scope or turned out to be systems which are huge and difficult to manage.

The problems connected with maintenance increase with the size of the application. For big tasks initially not all requirements are known and some may change over time. In addition, the understanding of the task increases which results in the desire to improve the system. As a consequence, one wants highly flexible approaches that allow continuous changes in the system. This in principle again favors the declarative character of knowledge-based systems. However, as indicated above, the successes have been somewhat limited. In Sect. 5 we will mention some recent developments which give reason for a new optimism.

3 Basic Ways of Knowledge Representation

A major distinction between pieces of information is whether they are regarded as either clearly true or clearly false, or whether this is not completely clear. In this section we will deal with the first situation; uncertain knowledge is treated in Sect. 4.

The dichotomy true/false is the viewpoint of classical logic and therefore it is no surprise that the representation languages are oriented towards classical predicate logic with its model theoretic semantics. In principle, all of predicate logic could be used. The problem is that the logical inference mechanisms are too inefficient for practical purposes. A basic observation now is that the restriction to fragments of predicate logic is helpful; the addition of some nonlogical features may again improve the efficiency. This leads to a rich diversity of possibilities.

The methods and styles of representing knowledge are of rather different character and we will discuss the major types used. The keyword for the first style is *rule-based programming*. A rule is of the form

$$A_1, \ldots, A_n \to B$$

Here the A_i and the B are formulas of the first order predicate calculus, the A_i are premises and B is the conclusion. Such a rule can be used in two ways:

- forward chaining: if the system knows that $A_1, \ldots, A_n$ are true, then the truth of B can be inferred.
- backward chaining: in order to confirm the truth of B, one needs only to establish the truth of $A_1, \ldots, A_n$.

As a generalization, B can also denote an action which can be carried out if the premises are true; in this case only forward chaining applies.

In addition to such rules one has also facts A which are considered to be unconditionally true. Variables may be involved which necessitate substitutions in order to unify the formulas in such a way that the rules can be applied as in the propositional case.

The first generation of AI programming languages was based on such rules. Backward chaining was realized in PROLOG and forward chaining has led to the development of OPS 5. Both languages (and their extensions) have been successfully used to implement expert systems. Although a rule is similar to a classical IF-THEN statement, rule-based programming may have significant advantages over traditional procedural programming. Firstly, the development of the program may be much faster because one has to deal with fewer implementation details. Secondly, a small change in the knowledge leads only to the replacement of a few rules and not to the rewriting of the whole program. A third aspect is that an explanation of the result is much easier than in a conventional algorithm: To know which rules have been applied very often gives sufficient information of how the result was obtained.

Rule-based programming, however, may also have serious drawbacks. Although it was just claimed as an advantage that one need not to be bother with procedural details, one might also in certain situations regret the lack of such a possibility. This is the case, for instance,when one knows an optimal algorithm and does not want to depend on the inference engine. This leads to the desire to incorporate procedural elements into a declarative program. Another missing element from rule systems is the principle of modularization and data abstraction. A third and important weakness is that there is no structure on the predicates. One cannot express easily that one predicate is more general than another, or that some predicate is an instance of a general concept.

Rules can be regarded as logical formulas which are "applied in a certain direction" (i.e. forward or backward,). Logical formulas without such a favored direction are called constraints. Such a set of constraints defines a Constraint Satisfaction Problem (CSP); a solution of the CSP is an assignment to the variables such that all constraints become true. The technique for solving CSPs is called constraint satisfaction; this has led to very sophisticated methods and constitutes the second style in our discussion. Recent developments have learned from practical applications that CPSs are often formally unsolvable. This is due to the fact that the constraints reflect contradicting demands. The crucial point is that these demands are not equally important, a fact which has led to the technique of constraint relaxation where less important constraints are stepwise removed (cf. Meyer (1994)). This introduces an aspect of optimality into the discussion: Some solutions may be better than others because they violate fewer or less important constraints.

A third less widely known way of making predicate logic operational runs under the name "terminological logic". It has only unary and binary formulas, called concepts and roles. Starting with a fixed number of concepts and roles news concepts can be defined is the following way:

if $\varphi(x)$ is a concept and $R(x,y)$ is a role then both
$\exists y \ (R(x,y) \wedge \varphi \ (y))$ and
$\forall y \ (R(x,y) \rightarrow \varphi \ (y))$
are concepts.

If for example, $\varphi(x)$ means "x is male" and $R(x,y)$ means "x is father of y" then one can define the concepts "x is the father of a son" and "x has only sons as children".

As the name indicates, terminological logic is in useful particular when precise definitions play a role. The inference mechanism can among others answer questions like

"Is concept φ more general then concept Ψ ?"

The "more general" relation gives rise to hierarchy (the "subsumption hierarchy"). Such hierarchies also dominate the object-oriented approach, see below. There are various implementations of terminological logic; an important one is KL-ONE. It is interesting to note that KL-ONE was first designed as a system independent of logic; the logical aspects were only detected later on. (cf. Brachmann and Schmolze (1985), Hanschke (1993)).

Next we come to common programming paradigms which are widely used in classical programming as well. As a fourth style of declarative programming we mention *functional programming*. In functional programming each object is a function, and hence the distinction between a function and an argument disappears. In particular, a function can be applied to itself. A functional expression, usually consisting of a function and given arguments is converted into a certain value and the only aspect one is interested in is the relation between a functional expression and its value. This corresponds very much to the pure mathematical treatment of functions and functional programming can always be regarded as a certain kind of recursion theory or, equivalently, an implementation of the λ-calculus.

The fifth style is the so-called *object-oriented programming style*. Objects are data capsules which have an inner life invisible to the other objects. An object has data and methods and the data belong to the inner life, i.e., they are invisible. The programming paradigm consists of a communication model; this means that objects have to send messages to each other in order to apply certain methods to certain arguments. If one has several objects with the same operations and a data space of the same structure then it is natural to form the class of such objects. It is then possible to define the common properties of those objects alone which are members of this class. Individual objects can then be considered as instances of that class. In the sense of this theory this realizes the membership relation. The second natural step is to realize the subset relation. This leads to a hierarchy of classes which can be represented by a directed graph. An edge leads from one class to another and points to the more special class. In principle, we encountered such hierarchies when we discussed terminological logic. Figure 2 shows such a hierarchy.

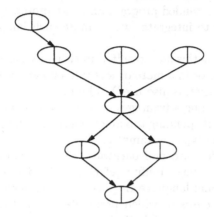

Fig. 2.

Such a hierarchy allows one to store each method or operation only once in the more general object. The concept of *inheritance* means that each piece of information is transferred automatically from the superclass to a subclass if it is needed there. A programming language entirely based on the object and class principle is SMALLTALK 80. In addition, Fig. 2 reflects the division of smalltalk objects into data and methods.

These programming styles are the basis for more complex methods of knowledge representation. The most important concept in this context is that of a *frame*. A frame is a structured object which has some similarity with the RECORD of PASCAL. It can be considered as a complex structured variable which has various *slots* that can be *filled* with values. The values may again be structured; in particular they may be frames themselves. Other possibilities are that slots are filled with functions, rules or simply unstructured variables. In the same way as objects are arranged in an inheritance hierarchy, this can be done with frames too. There can, however, be various other edges defined between frames. An example is the *connected-to-edge* if the frames are considered to represent components of a technical system.

It is, of course, desirable to define frames in a way that is as expressive and comfortable as possible. This may, however, lead to serious implementation problems. Modern knowledge representation tool kits almost always contain, besides rules and functions, a certain concept of a frame. In most cases these systems are hybrid, which means that rules, functions, and frames are in different packages. This creates problems when one task requires the combined use of such packages because of the necessary communication overhead.

Logic, functional, and object-oriented programming and frames are appropriate for the representation of different kinds of information. In reality these types of information cannot be clearly separated. Therefore, after having chosen a specific programming style, this often turns out to be suitable for some parts of the intended program and less appropriate for others. This has led to the desire to integrate two or more such languages into a single one.

Attempts to achieve this have resulted in either extensions of one language by features of another or in a more or less full amalgamation (in this context the term deep integration is also used). The problem was always to improve one language by including advantages of a second one without importing its drawbacks and without spending too much effort in managing the integration. We will mention some typical examples.

A integration of forward and backward chaining is described in Hinkelmann (1995). The language Relfun (cf Boley (1996)) contains both PROLOG and the functional language Lisp as a sublanguage by introducing the concepts of *valued clauses* to which predicates as well as functions can be reduced. The extension of PROLOG by constraints has led to various systems of constrained logic programming (see, e.g., Dincbas, Simony and Hentensyck (1990)) which have proven to be commercially successful. The languages TAXON and TAXLOG combine terminological logic with constraints and with PROLOG, respectively (cf. Abecker and Wache (1994)).

Originally logic-oriented and object-oriented languages were only loosely coupled. Later on several proposals for a more intensive integration of the two formalisms were made (cf. Ait-Kaci and Nasser (1986)). In principle one starts with a predicate calculus with several sorts such that these sorts are ordered by inclusion. The sorts (see below) correspond to (and are identified with) the classes in object-oriented programming. In the realization of extensions of PROLOG one can basically observe three levels in the density of the integration:

- coupling of PROLOG with an object-oriented language
- implementation of an object system in PROLOG
- extension of PROLOG by sorts.

The first approach consists mainly of the definition of a suitable interface of PROLOG with, e.g., SMALLTALK. In principle we still have two independent processes which can call each other. In order to implement an object system in PROLOG one defines additional PROLOG predicates, which enable the system to generate objects and to activate methods by messages. For an object-oriented extension of PROLOG the most plausible approach is to use sorts. Sorts are attached to variables, constants and arbitrary terms. Instead of formulas $P(x, y)$ one can build formulas like $P(x_S, y_T)$ in sorted logic. The meaning is "P holds for x, y_T where x_S is of sort S and y_T is of Sort T'''. The interpretation of the sorts is a certain subset of the model. An example is the expression "x_S in married to y_T" where S is the sort of male objects and T is the sort of female objects. Sorted logic can easily be implemented in PROLOG. Sorts are represented by unary predicates and inheritance corresponds to certain implications. An improvement over this simple approach uses mainly two techniques:

- Inference steps which correspond to a search in the class hierarchy are built in to the unification algorithm.
- In answer substitutions sorted variables may occur. This avoids an enumeration of all individuals via backtracking.

A very general approach to the integration problem was done in the CO-LAB system developed at the DFKI Kaiserslautern, (cf. Boley et al. (1995)). COLAB integrates, to various degrees, forward and backward chaining, functional programming, constraints and terminological logic.

4   Vague and Uncertain Knowledge

The representation techniques discussed so far assume tacitly the view of classical logic that everything is either true or false. In most applications this assumption is frequently not justified. In order to deal with such situations one has first to distinguish different notions of uncertainty. They all have in common that besides "true" and "false" concepts of "approximation of truth" are introduced. Such concepts usually result in additional truth values. Their meaning depends on the type of uncertainty one considers.

(a) The probability-based approach considers propositions of essentially two-valued logic where one does not have not enough information available to tell which truth value is obtained. The uncertainty is described by a probability about the truth of the proposition. This whole area falls in principle into the domain of probability theory and statistics and has been integrated into knowledge-based systems in various ways. One major point is to extend the inference system in order to get the propagation of probabilities under control.

A particular problem arises when there is not even enough information to yield a probability distribution. One way is to use evidences which assign probabilities to statements like "The true statement is contained in a certain set S of statements". If one chooses S as the set of all statements then this reflects total ignorance (cf. Shafer (1976)). Another possibility is to introduce intervals for probabilities:

$$0 \leq \alpha \leq \text{Prob } (p \text{ is true}) \leq \beta \leq 1$$

Here the choice of $\alpha = 0$ and $\beta = 1$ reflects total ignorance (cf. Weichselberger and Pöhlmann (1990)).

(b) The fuzzy approach considers statements which by their linguistic nature have no precise truth value. Two cases can be distinguished:

(b1) The fuzzy statement is an abstraction of a precise statement like "patient X has high fever". A fuzzy function μ would assign a real number to this statement reflecting the subjective degree to which one would assume high fever. What happens is that the predicate

high-fever(X))

no longer defines a subset $H \subseteq P$ (P set of all patients, H set of patients with high fever) but a membership function

$$\mu_H : P \rightarrow [0,1]$$

where $\mu_H(X)$ is the "degree of membership" of X to H.

(b2) In the second situation one considers statements like "patient X is pale". There exists no original value of "paleness" from which it is abstracted, the statement is totally subjective. The technique of generalized membership functions works here too. Also the fuzzy technique has been used extensively in knowledge-based systems.

5 Episodic Knowledge and Case-Based Reasoning

For all the techniques described so far it is necessary to have the knowledge described in a form suitable for the representation formalism in question. Very often this is not the case and the knowledge acquisition techniques have to transform the knowledge from the available knowledge sources into the required format as indicated in Fig.1. This is the so-called "knowledge acquisition bottleneck" which has been a major obstacle in creating successful expert systems: Knowledge acquisition is often simply too costly to perform.

One way to overcome this problem is the development of special knowledge acquisition systems. They support the systems developer by carrying out the A-steps of Fig. 1 in a stepwise controlled and carefully protocolled way. This technique also supports the maintenance of the system when the context changes.

A very comprehensive approach is known under the name KADS ("Knowledge Acquisition Document System"), see, e.g., Wielinga et al. (1990). In KADS the idea is to organize and partly formalize the knowledge acquisition process in such a way that it can be supported by machines. KADS distinguishes two main models: The conceptual model and the design model. These are not models in the sense of logic but they describe two main phases in the acquisition process. The conceptual model is an assembly of all concepts of interest for the problem in question; the design model describes the way which is selected to solve the problem. Special implemented systems are described in Schmidt (1995) and Maurer (1993). A typical example of knowledge that is difficult to formalize is when the sources describe problems of the intended kind and their solutions as documented in the past, e.g., records of patients together with their diseases. We call such past events *episodes* and their descriptions *episodic knowledge*. Such episodic knowledge contains the knowledge in question only implicitly. Instead of undergoing the time consuming task of making it explicit the idea of case-based reasoning is to use it directly. If one considers the behavior of humans this looks very natural because they often solve problems using the past experience of solving similar problems.

One first needs a definition of what a case is. In its simplest version a case is an ordered pair

$$C = (P, S)$$

where P is a problem description and S is a solution description. A case base is a set CB of cases. The central notion of case-based reasoning is a concept of *similarity* of problem descriptions. Similarity is obviously not a classical binary predicate but again introduces a degree of uncertainty and a notion of "more or less similar". A formalization leads to numerical function in form of a similarity measure

$$\text{sim}(P1, P2) \in [\,0, 1\,]$$

With such a measure one can define the concept of a nearest neighbor of some P with respect to CB:

$$NN(P, CB) = Q$$
$$\Leftrightarrow$$
$(Q, S) \in CB$ for some S and
for all $(Q', S') \in CB$ $\text{sim}(Q',P) \leq \text{sim}(Q, P)$ holds.

If we regard the case base as a treasure of previous experience the nearest neighbor of an actual problem should represent the most useful experience in the case base.

Now we can describe the principle by which case-based reasoning proceeds for an actual problem P:

(1) Determine some $Q = NN\ (P, CB)$, $(Q, S) \in CB$
(2) Take S as a solution for P too or apply some transformation T
     to S and take $T(S)$ as a solution for P.

This reflects human behavior: Take the solution from the "most similar" problem directly or modify it appropriately. T in (2) is called the *solution transformation*. In many situations it is an obvious adaption which reflects the major differences between P and Q. For example, if patient Q had a certain desease in the right ear and patient P has exactly the some symptoms only in the left ear, then T would simply replace "right ear" in Q's diagnosis by "left ear" in order to obtain a diagnosis for P.

This problem solving method also requires certain prerequisites in order to work well:

(1) Sufficiently many cases are recorded.
(2) There is a suitable notion of similarity.
(3) The solution transformation is available.

There are many examples where these prerequisites are satisfied and case-based reasoning has been successfully applied. Intuitively this reflects the observation that it is often easier to tell that two problems are similar than it is to exploit the details of the solution. It should be emphasized that the most crucial notion is that of similarity. The similarity of two objects is not of an absolute character but depends on the particular problem situation. In some sense we encounter a circular definition:

Two problems are similar if their solutions are similar.

The point is that we assume that there is a notion of "similarity of solutions". In the simplest situation this means that the solutions are identical; in more complicated situations it can mean that the solutions can easily be transformed into each other. The problem in the definition of a similarity measure is to trace the similarity of solutions back to the problems. In the light of this it is safe to say that a *well-informed measure "knows much"* about the problem.

If the representation of the objects is given by values of certain attributes then each is a vector $a = (a_1, ..., a_n)$ where the a_i are values of certain attributes A_i. A very simple object similarity measure is given by

$$H(a, b) = |\{ i \mid a_i = b_i \}|$$

which is the well known Hamming similarity. As an improvement one can reflect the importance of attributes using a weight vector $g = (g_1, ..., g_n)$ of real numbers. This leads to the weighted Hamming similarity

$$H_g(a, b) = \Sigma (g_i \mid a_i = b_i).$$

This approach has been further refined; in particular, representations have been allowed where the attribute values are partially unknown.

The approach described so far circumvents many problems of knowledge acquisition. In terms of computer science, the episodic knowledge in the cases is not compiled into rules, constraints, etc., but interpreted at run time. Hence the compilation costs are saved. There is, however, a price to pay because searching in the case base may be very time consuming too. But here a second aspect of case-based reasoning enters the picture: Better understanding of the problem type over time can lead to an improvement of the system by improving the similarity measure and removing unnecessary cases; the advantage is that most of the main features of the system remain unchanged. For this reason case-based reasoning can be regarded as a logical step forward in the development of systems which meets the requirements of flexibility coming from software engineering as outlined in the introduction.

A further step is to generalize the notion of a case. The features of case-based reasoning by no means require that one deals with problems and their exact solutions. Instead one can take

P: A general problematic situation

S: A hint for a solution or even an arbitrary piece of useful information

Hence we discover here some aspects of information retrieval which make the technique particularly suitable for interactive working.

In summary, the technique of using past cases is a new development and seems to have some advantages when traditional knowledge-based systems have difficulties. A current research topic is to elaborate this relation in more detail. This should also be a part of the enterprise to get a better picture of how knowledge-based systems are related to general problem-solving methods that have been developed in computer science over years (cf. Weß, Althoff and Richter (1994)).

6 What Kind of Knowledge Has Been Represented?

Most applications of knowledge-based systems fall into one of the categories diagnosis, configuration, and planning and are found either in a technical area or in the domain of medicine. Therefore it is no surprise that the knowledge bases describe

- properties of the domain of interest, e.g., knowledge about machines or diseases
- heuristics about the suggested behavior in a planning or diagnostic situation.

The description of the domain of interest is called a model and reasoning using the model is called the model-based approach. If only correct logical deductions are employed only correct conclusions can be drawn, e.g. a fault diagnosis obtained this way will always be correct. However, this way of drawing conclusions may be very inefficient. This is due to the fact that special and useful pieces of advice which an expert (informally) knows are not represented in the model. Here the heuristics enter the discussion. Most of the heuristics are of the form

"If A is true or plausible then B should be considered."

Hence heuristics connects pieces of information not in a logical form (although it is represented as a rule) but as an association which is useful. Therefore one talks in this context about "associative" knowledge.

An intermediate form between associations and models is the use of taxonomies (or hierarchies). Taxonomies bring structure into associative knowledge but are not as complete and explanatory as models.

According to this we can distinguish three classical phases in the history of knowledge representation: associations, taxonomies, and models. We now already have a next phase where similarities are used. Yet another phase of increasing importance is connectionism which will, however, not be considered here (but see Richter (1994)).

The three phases mentioned have an interesting historical counterpart. They have been described by M. Foucault in his book 'Les mots et les choses'. He describes the development of science from the 17th to the 19th century and identified three periods. He does not give a historical description in the usual sense but a systematic approach to discover common methods in different disciplines. These common elements can be regarded as fundamental codes of a culture; at the corresponding times no-one was explicitly aware of them.

' In the first phase thinking was dominated by what Foucault called *similarities*, today one would call it *associative knowledge*. A typical piece of knowledge of that form was a relation between a certain disease and its medical treatment. Such a relation was not explained and therefore in some sense of a magic character. The world seemed to be an accumulation of such knowledge units and the aim of research was to collect as much knowledge as possible in this way.

Nevertheless there was a desire to structure the knowledge and this led to the second phase which can be characterized by the term *taxonomia*. The main task of research was to structure and to classify knowledge. A prominent example is the classifica One was still not able to explain the underlying reasons for observed behavior but an important step was made in this direction. The third and Foucault's final period can be described by the term *modelling*. The intention of a model is to describe a

IF interval X overlaps interval Y from the left

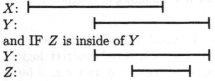

and IF Z is inside of Y

What are the possible relations between X and Z?

The study of time in this way can also be considered as an investigation of a one-dimensional space. A much more difficult problem is the qualitative behavior of 2- or 3-dimensional spatial relations. This is currently the subject of intensive research involving computer scientists, psychologists, and cognitive scientists.

The description of spatial relations is important in everyday life and part of common sense. A natural question that arises is which other parts of common sense can be represented in a knowledge base. This problem was tackled in the CYC-Project (see Lenat and Guha (1990)) which is probably the most ambitious project in the history of AI. The original idea was to formalize *all* the common sense knowledge a person has. Isn't that too much?

Suppose a person learns each day about 30 000 new things for a period of 30 years; this would amount to about 100 million entries into the knowledge base. The attempt to collect such a huge set of information was later on revised and replaced by quite a number (several hundreds) of *microtheories* each focusing on a particular domain of knowledge.

7 Concluding Remark

In summary we can state that knowledge representation is still an area of very active research. New methods have been discovered and older ones have been improved. The possibility of representing knowledge formally has also led to a better structuring of many areas where previously there was no need for precise structures. Hence advances in knowledge representation are of increasing importance to non-computer scientists.

References

Abecker A., Wache H.: A Layer Architecture for the Integration of Rules, Inheritance, and Constraints. ICLP 1994 Post-Conference Workshop on the Integration of Declarative Paradigms, ed. Ait-Kaci et al.

Ait-Kaci H., Nasser R.: LOGIN: Electronic Programming Language with Built in Inheritance. J. of Logic Programming 3 (1986), p. 185–215

Allen J.F.: Towards a General Theory of Action and Time. Artificial Intelligence 23 (1984), p. 123–154

Boley H., Hanschke Ph., Hinkelmann K., Meyer M.: COLAB: A hybrid knowledgerepresentation and compilation laboratory. Annals of Oper. Res. 55 (1995), p. 11–79

Boley H.: A Tight, Practical Integration of Relations and Functions. Habilitationsschrift Kaiserslautern 1996

Brachmann R.J., Schmolze J.G.: An overview of the KL-ONE representation System, Cognitive Science 9 (1985), p. 171–216

Dincbas M., Simony A., v. Hentensyck P.: Solving large combinatorical problems in logic programming. Journal on Logic Programming 8 (1990), p. 75–93

Foucault M.: Les Mots et les Choses. (Editions Gallimard, 1966).

Hanschke Ph.: A Declarative Integration of Terminological, Constraint-based, Data driven and Goul-directed Reasoning. Ph. Thesis Kaiserslautern 1993 (also infix-Verlag 1996).

Hinkelmann K.: Transformation von Hornklausel-Wissensbasen: Verarbeitung gleichen Wissens durch verschiedene Inferenzen, (infix-Verlag, 1995).

Lenat D., Guha R.V.: Building Large Knowledge-Based Systems. (Addison-Wesley, 1990).

Maurer F.: Hypermediabasiertes Knowledge Engineering für verteilte wissensbasierte, (infix-Verlag, 1993).

Meyer M.: Finite Domain Constraints: Declarativity meets Efficiency, Theory meets Application, (infix-Verlag, 1994).

Richter M.M.: Self-Organization, Artificial Intelligence and Connectionism. In: On Self-Organization, ed. Mishra R.K., Maa D., Zwierlein E. (Springer-Verlag, 1994), p. 80–91

Shafer G.: A mathematical theory of evidence. (Princeton University Press, 1976).

Schmidt G.: Modellbasierte, interaktive Wissensakquisition und Dokumentation von Domänenwissen MIKADO), (infix-Verlag, 1995).

Weichselberger K., Pöhlmann S.: A Methodology for Uncertainty in Knowledge-Based Systems. Springer Lecture Notes in AI 419 (1990).

Weß S., Althoff K.D., Richter M.M.: Topics in Case-Based Reasoning. EWCBR'93, Selected Papers, Springer LNAI 837 (1994).

Wielinga B., Boose, B. G. Gaines, G. Schreiber, M. van Someren (eds.): Current Trends Knowledge Acquisition (IOS Press, 1990).

Neural Networks

Heinz Horner and Reimer Kühn

Institut für Theoretische Physik, Universität Heidelberg,
Philosophenweg 19, D-69120 Heidelberg, Germany.
e-mail: horner@tphys.uni-heidelberg.de und kuehn@tphys.uni-heidelberg.de

1 Introduction

Understanding at least some of the functioning of our own brain is certainly an extraordinary scientific and intellectual challenge and it requires the combined effort of many different disciplines. Each individual group can grasp only a limited set of aspects, but its particular methods, questions and results can influence, stimulate and hopefully enrich the thoughts of others. This is the frame in which the following contribution, written by theoretical physicists, should be seen.

Statistical physics usually deals with large collections of similar or identical building blocks, making up a gas, a liquid, or a solid. For the collective behavior of such an assembly most of the properties of the individual elements are only of marginal relevance. This allows one to construct crude and simplified models which nevertheless reproduce certain aspects with extremely high accuracy. An essential part of this modeling is to find out which of the properties of the elements are relevant and what kind of questions can or cannot be treated by such models. The usual goal is to construct models that are as simple as possible and to leave out as many details as possible, even if they are perfectly well known. The natural hope is that the essential properties can be understood better on a simple model. This, on the other hand, seems to contradict the ideals of modeling in other disciplines and this can severely obstruct the interdisciplinary exchange of thoughts.

Our brain (Braitenberg and Schüz, 1991) is certainly not an unstructured collection of identical neurons. It consists of various areas performing special tasks and communicating along specific pathways. Even on a smaller scale it is organized into layers and columns. Nevertheless the overwhelming majority of neurons in our brain belong to one of perhaps three types. Furthermore, on an even smaller scale, neurons seem to interact in a rather disordered fashion, and the pathways between different areas are to some degree diffuse. Keeping in mind that models of neural networks with no a priori structure are certainly limited, it is of interest to see how structures can evolve by learning processes and what kind of tasks they can perform.

Over the last ten or more years, abstract and simplified models of brain functions have been a target of research in statistical physics (Amit, 1989;

Hertz et al., 1991). A model of an associative memory was proposed by Hopfield in 1982, following earlier work by Caianello and Little. This model is not only based on extremely simplified neurons (McCulloch-Pitts, 1943), it also serves a heavily schematized task, the storage and retrieval of uncorrelated random patterns. This twofold idealization made it, however, tractable and the source of quantitative results. In the meantime there have been many extensions of this model, some of which will be discussed later on. One of the essential points of this model is the fact that information is stored in a distributed fashion in the synaptic connections among the neurons. Each synapse carries information about each pattern stored, such that destruction of part of the synapses does not destroy the whole memory. The storage of a pattern requires a learning process which results in a modification of the strength of all synapses. The original model was based on a simple learning rule, essentially the one proposed by Hebb, which is in a sense a neuronal manifestation of Pavlov's ideas of conditioned reflexes. Regarding learning, more sophisticated rules have also been investigated and are discussed later.

Even restricting ourselves to this kind of models, we can sketch only a small part of what has been worked out in the past, and also only small parts of the progress in getting those models closer to biology. It is interesting to note that artificial neural nets, in the form of algorithms or hardware, have found many technical applications. This aspect, however, will be left aside almost completely.

Before entering the discussion of learning or memory, we want to give a brief overview over the biological background on neurons, their basic functioning, and their arrangement in the brain (Braitenberg and Schüz, 1991; Abeles, 1991).

2 Biological Versus Formal Neural Nets

2.1 Biological Background

A typical neuron, e.g. a pyramidal cell (Fig. 1), consists of the cell body or soma; extending from it there is a branched structure of about 2 mm diameter, called the dendritic tree, and the nerve fiber or axon, which again branches and can have extensions reaching distant parts of the brain. The branches of the axon end at so-called synapses which make contact to the dendrites of other neurons. There are of course also axons coming in from sensory organs or axons reaching out to the motor system. Compared to the number of connections within the brain, their number is very small. This amazing fact suggests that the brain is primarily busy analyzing the sparse input or shuffling around internal information.

The main purpose of a neuron is to receive signals from other neurons, to process the signals and finally to send signals again to other cells. What happens in more detail is the following. Assume a cell is excited, which means that the electrical potential across its membrane exceeds some threshold. This creates a short electric pulse, of about 1 ms duration, which travels along the axon and ultimately reaches the synapses at the ends of its branches. Having sent a spike, the cell returns to its resting state. A spike arriving at a synapse releases a certain quantity of so-called neuro–transmitter molecules which diffuse across the small gap between synapse and dendrite of some other cell. The neuro–transmitters themselves then open certain channel proteins in the membrane of the postsynaptic cell and this finally influences the electrical potential across the membrane of this cell. The neuro–transmitters released from pyramidal cells have the effect of driving the potential of the postsy-naptic cell towards the threshold, their synapses are called excitatory. There are, however, also inhibitory cells with neuro–transmitters having the oppo-site effect. The individual changes of the potential caused by the spikes of the presynaptic cells are collected over a period of about 10 ms and if the threshold is reached the postsynaptic cell itself fires a spike. Typically 100 incoming spikes within this period are necessary to reach this state.

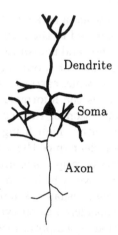

Fig. 1. Schematic view of a neuron

The human brain contains 10^{10} to 10^{11} neurons and more than 10^{14} synap-tic connections among them. The neurons are arranged in a thin layer of about 2 mm on the brain's outer surface, the cortex, and each mm^2 contains typically 10^5 such cells. This means that the dendritic trees of these cells penetrate each other and form a dense web. Part of the axons of these cells again project onto the dendrites in the immediate neighborhood and only a fraction reaches more distant regions of the brain. This means that on a scale of a few mm^3 more than 10^5 neurons are tightly connected. This does not imply that more distant regions are weakly coupled. The huge amount of white matter containing axons connecting more distant parts only indicates the possibility of strong interactions of such regions as well.

It is tempting to compare this with structures which we find within the integrated circuits of an electronic computer. The typical size of a synapse is 0.1 μm, whereas the smallest structures found in integrated circuits are about five times as big. The packing density of synapses attached to a dendrite is about 10^9 per mm^2, whereas only 1/1000 of this packing density is reached in electronic devices. A comparison of the computational power is also impressive. A modern computer can perform up to 10^9 elementary operations per second. The computational power of a single neuron is quite low, but they all work in parallel. Assuming that a neuron fires at a rate of 10 spikes per second, which is typical, and assuming that each spike transmitted through a synapse corresponds to an elementary computation, we find a computing power of about 10^{15} operations per second. These numbers have to be kept in mind when we try to imitate brain functions with artificial devices.

2.2   Formal Neurons – Spikes Versus Rates

The actual processes going on when a spike is formed or when it arrives at a synapse and its signal is transmitted to the next neuron, involve an interplay of various channels, ionic currents, and transmitter molecules. This should not be of concern as long as we are interested only in the data processing aspects. A serious question, however, is what carries the information? Is it a single spike and its precise timing or is the information coded in the firing rates? For sensory neurons the proposition of rate coding seems well established. These neurons typically have rather high firing rates in their excited state. For the brain this is much less clear since the typical spike rates are low and the intervals between two successive spikes emitted by a neuron are longer or at best of the order of the time over which incoming spikes are accumulated. Nevertheless a spike rate coding is usually assumed for the brain as well. This means that a rate has to be considered as an average over the spikes of many presynaptic neurons rather than a temporal average over the spikes emitted by a single cell. This is plausible, having in mind that typically 100 or more arriving spikes are necessary to release an outgoing spike. This suggests for the firing rate ν_i of a neuron i

$$\nu_i(t) = \Phi\left(\sum_j W_{ij}\nu_j(t) - \vartheta\right) , \tag{1}$$

where ϑ is some threshold and $\Phi(x)$ is some increasing function of x. The simplest assumption is $\Phi(x) = 0$ for $x < 0$ and $\Phi(x) = 1$ for $x > 0$. This model was first proposed by McCulloch and Pitts (1943). The quantity W_{ij} describes the coupling efficacy of the synapse connecting the presynaptic cell j with the postsynaptic cell i. For excitatory synapses $W_{ij} > 0$, and for inhibitory synapses $W_{ij} < 0$.

This model certainly leaves out many effects. For instance the assumption of linear superposition of incoming signals neglects any dependence on the position of the synapse on the dendritic tree.

Investigating networks of spiking neurons, researchers have again designed simplified models. One of them is the integrate and fire neuron which mimics the mechanism of spike generation at least in a crude way. It sums up incoming signals by changing the membrane potential. As soon as a certain threshold is reached, the neuron fires and the membrane potential is reset to its resting value.

If rate coding is appropriate, results obtained for the first kind of networks should be reproduced by networks of spiking neurons as well. On the other hand there are many questions which can only be taken up within the framework of spiking neurons, for instance which role the precise timing of spikes plays (Abeles, 1991) or whether the activity in a network with excitatory couplings between its neurons can be stabilized by adding inhibitory neurons (Amit et al., 1994, 1997).

2.3   Hebbian Learning – Sparse Coding

The most remarkable feature of neural networks is their ability to learn. This is attributed to a certain plasticity of the synaptic coupling strengths. The question is of course, how is this plasticity used in a meaningful way?

The basic idea proposed by Hebb (1949) actually goes back to the notion of conditioned reflexes put forward by Pavlov. Assume a stimulus A results in a reaction R. If simultaneously with A a second stimulus B is applied, then after some training stimulus B alone will be sufficient to trigger reaction R, although this was not the case before training. Let A be represented by the activity of a neuron ℓ and the reaction R by neuron i becoming active. This would be the case if the coupling $W_{i\ell}$ is sufficiently strong. Before training, the stimulus B, represented by the activity of neuron j, is assumed *not* to trigger the reaction R. That is, the coupling W_{ij} is assumed to be weak. During training with A *and* B present, the cells j and i are simultaneously active, the latter being activated by cell ℓ. Assume now that the synaptic strength W_{ij} between neurons j and i is increased if both cells are simultaneously active. Then, after some training, this coupling W_{ij} will be strong enough to sustain the reaction R without A being applied, provided B is present. This is represented by the Hebb learning rule

$$\Delta W_{ij} \propto \nu_i \, \nu_j \, . \tag{2}$$

Most remarkably this learning rule does not require a direct connection between the cells ℓ and j representing the stimuli A and B. That is, the equivalence of stimuli A and B has been learnt without any a priori relation between A and B. What has been used is only the simultaneous occurrence of A and B. Despite its simplicity this learning rule is extremely powerful.

It is not completely clear how such a change in the synaptic efficacies is realized in detail, whether it is caused by changes in the synapse itself or by changes in the density of receptor proteins on the membrane of the dendrite of the postsynaptic neuron. Nevertheless it is plausible at least in the sense that this learning process depends only on the simultaneous state of the pre- and postsynaptic cell. It is generally assumed that learning takes place on a time scale much slower than the intrinsic time scale of a few milliseconds characteristic of neural dynamics.

We can now go one step further and consider the learning of more than one pattern. A pattern is a certain configuration of active and inactive neurons. A pattern, say μ, is represented by a set of variables ξ_i^μ for each pattern and neuron. This means that in pattern μ neuron i fires with a rate ξ_i^μ. In the most simple case $\xi_i^\mu = 1$ if i is excited and $\xi_i^\mu = 0$ otherwise. Having learnt a set of patterns the couplings, according to the above learning rule, have the values

$$W_{ij} = W_0 \sum_\mu \xi_i^\mu \xi_j^\mu \tag{3}$$

where W_0 has the meaning of a learning strength. Actually this learning strength might also depend on the kind of pattern presented, for instance on whether the pattern is new, unexpected, relevant in some sense or under which global situation, attention, laziness or stress, it is presented. This can lead to improved learning or suppression of uninteresting information.

The above learning rule is constructed such that, at least for $W_0 > 0$, only excitatory couplings are generated. This is in accordance with the finding that the pyramidal neurons have excitatory synapses only and that the plasticity of the synapses is most pronounced in this cell type. On the other hand, this causes a problem. A network with excitatory synapses only would quickly go into a state where all neurons are firing at a high rate. The cortex, however, contains inhibitory cells as well. The likely purpose of these cells is to control the mean activity of the network and to prevent it from reaching the unwanted state of uniform high activity. A malfunctioning of this regulation is probably the cause of epileptic seizures.

Actually the mean activity in our brain seems to be rather low. This means that at a given time only a small percentage of the neurons is firing at an elevated rate. Typical patterns are sparse, having many more 0's than 1's. This is a bit surprising since the maximal information per pattern is contained in binary patterns with approximately equal number of 0's and 1's and such symmetric coding is also used in our computers. Nevertheless there are several good reasons for sparse coding, some of which will be discussed later.

In the original Hopfield model the degree of abstraction is pushed a step further. Here symmetric patterns with equal number of 0's and 1's are considered. This requires a modified learning rule. First of all the inactive state is now represented by -1 rather than 0. With this modification the above learning rule can again be used, but now the coupling strength can also be weakened and the couplings can acquire negative values. Furthermore it is assumed that each neuron is connected to every other neuron and that the couplings between two neurons have the same value in both directions. This is certainly rather unrealistic in view of the biological background. Modified models with one or the other simplifying assumption removed have been investigated as well. However, they show quite similar behavior. This demonstrates the robustness of the models with respect to modifications of details, which again serves as a justification for this simple kind of modeling.

2.4   Transmission Delays

The propagation of a spike along the axon, the transmission of this signal across the synapse, and the propagation along the dendrite take some time. This causes some total delay τ_{ij}, typically a few milliseconds, in the transmission of a signal from neuron j to neuron i. Incorporating this into Eq. (1) yields a modified form

$$\nu_i(t) = \Phi\left(\sum_j W_{ij}\nu_j(t - \tau_{ij}) - \vartheta\right). \tag{4}$$

As long as we are interested in slow processes, this delay is of no relevance. On the other hand it gives the opportunity to generate or learn sequences of patterns evolving in time. This might be of relevance in processes like speech generation or recognition, or in generating periodic or aperiodic motions. Other proposals use this mechanism for temporal linking of different features of the same object or for the segmentation of stimuli generated by unrelated objects. Another mechanism which might play a role in this context is the phenomenon of fatigue. This means that the firing rate of an excited neuron, even at constant input, goes down after a while. The associated time scales can vary from a few milliseconds up to minutes or hours. In any case there are several mechanisms which can be used for the generation or recognition of temporal structures and we shall return to this point later.

The picture developed so far certainly leaves out many interesting and important aspects. Nevertheless even this oversimplified frame allows us to understand some basic mechanisms. On the other hand it is far from being a description of the brain as a whole. What is certainly missing is the structure on a larger scale. In order to proceed in this direction one would have to construct modules performing special tasks, like data preprocessing or memory, and one would have to arrange for a meaningful interplay of those modules. This is currently far beyond our possibilities, as we are lacking analytic tools, computational power and, perhaps more importantly even, good questions and well formalizable tasks to be put to such a modular architecture.

3 Learning and Generalization

Given that we interpret the firing patterns of a neural network as representing information, neural dynamics must be regarded as a form of *information processing*. Moreover, disregarding the full complexity of the internal dynamics of single neurons, as we have good reasons to do (see Sect. 2.2), we find that the course of neural dynamics, hence information processing in a neural network, is determined by its synaptic organization.

Consequently, *shaping* the information processing capabilities of a neural network requires changing its synapses. In a neural setting, this process is called "learning", or "training", as opposed to "programming" in the context of symbolic computation. Indeed, as we have already indicated above, the process of learning is rather different from that of programming a computer. It is incremental, sometimes repetitive, and it proceeds by way of presenting "examples". The examples may represent associations to be implemented in the net. They may also be instances of some rule, and one of the reasons for excitement about neural networks is that they are able to *extract* rules from examples. That is, by a process of training on examples, they can be made to *behave* according to a set of rules which – while manifest in the examples – are usually never made explicit, and are quite often not known in algorithmic detail. Such is, incidentally, also the case with most skills humans possess (subconsciously). In what follows, we discuss the issues of learning and generalization in somewhat greater detail.

We start by analysing learning (and generalization) for a single threshold neuron, the perceptron. First, because it gives us the opportunity to discuss some of the concepts useful for a quantitative analysis of learning already in the simplest possible setting; second, because the simple perceptron can be regarded as the elementary building block of networks exhibiting more complicated architectures, and capable of solving more complicated tasks.

Regarding architectures, it is useful to distinguish between so-called feedforward nets, and networks with feedback loops (Fig. 2). In feedforward nets, the information flow is directed; at their output side, they produce a certain *map* or function of the firing patterns fed into their input layer. Given the

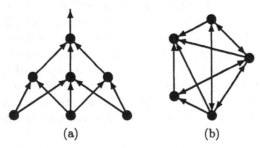

(a)   (b)

Fig. 2. (a) Feedforward network. (b) Network with feedback loops

architecture of such a layered net, the function it implements is determined by the values of the synaptic weights between its neurons. Networks with feedback loops, on the other hand, exhibit and utilize non-trivial *dynamical* properties. For them, the notion of (dynamical) attractor is of particular relevance, and learning aims at constructing desired attractors, be they fixed points, limit cycles or chaotic. We shall discuss attractor networks separately in Sect. 4. Finally, feedforward architectures may be combined with elements providing feedback loops in special ways to create so-called feature maps, which we also briefly describe.

The physics approach to analyzing learning and generalization has consisted in supplementing general considerations with quantitative analyses of heavily schematized situations. Main tools have been statistical analyses, which can however be quite forceful (and luckily often simple) when the size of a given information processing task becomes large in a sense to be specified below.

It goes without saying that this approach would not be complete without demonstrating – either theoretically, by way of simulations, or, by studying special examples – that the main functional features and trends seen in abstract statistical settings can survive the removal of a broad range of idealizations and simplifications, and that they, indeed, prove to be resilient against changing fine details at the microscopic level.

3.1 Simple Perceptrons

A perceptron mimics the functioning of a single (formal) neuron. Given an input $\nu = (\nu_1, \nu_2, \ldots, \nu_N)$ at its N afferent synapses, it evaluates its local field or post-synaptic potential as a weighted sum of the input components ν_j,

$$h_0(\nu) = \sum_{j=1}^{N} W_{0j} \nu_j \, , \tag{5}$$

compares this with a threshold ϑ, and produces an output ν_0 according to its transfer function or input–output relation

$$\nu_0 = \Phi(h_0(\nu) - \vartheta) \, . \tag{6}$$

For simple perceptrons, one usually assumes a step-like transfer function. Common choices are $\Phi(x) = \mathrm{sgn}(x)$ or $\Phi(x) = \Theta(x)$ depending on whether one chooses a ± 1 representation or a 1–0 representation for the active and inactive states[1].

The kind of functionality provided by a perceptron has a simple geometrical interpretation. Equation (6) shows that a perceptron implements a *two-class* classification, assigning an 'active' or an 'inactive' output-bit to each input pattern ν, according to whether it produces a super- or sub-threshold

---

[1] $\Theta(x)$ is Heaviside's step function: $\Theta(x) = 1$ for $x > 0$ and $\Theta(x) = 0$ otherwise.

local field. The dividing *decision surface* is given by the inputs for which
$\mathbf{W}_0 \cdot \boldsymbol{\nu} \equiv \sum_j W_{0j}\nu_j = \vartheta$. It is a *linear* hyperplane orthogonal to the direction
of the vector $\mathbf{W}_0$ of synaptic weights (W_{0j}) in the N-dimensional space of
inputs (Fig. 3). Pattern sets which are classifiable that way are called linearly
separable. The linearly separable family of problems is certainly non-trivial,
but obviously also of limited complexity. Taking Boolean functions of two inputs as an example, and choosing the representation $1 \equiv true$ and $0 \equiv false$,
one finds that $\text{AND}(\nu_1, \nu_2)$, and $\text{OR}(\nu_1, \nu_2)$ as well as $\text{IMPL}(\nu_1, \nu_2)$ are linearly separable, whereas $\text{XOR}(\nu_1, \nu_2)$ is not.

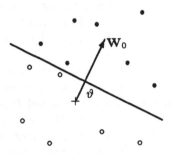

Fig. 3. Linear separation by a perceptron

It is interesting to see how Hebbian learning, the most prominent candidate for a biologically plausible learning algorithm, would perform on learning
a linearly separable set of associations. A problem that has been thoroughly
studied is that of learning *random* associations. That is, one is given a set of
input patterns $\boldsymbol{\xi}^\mu$, $\mu = 1, \ldots, P$, and their associated set of desired output
labels ζ_0^μ. Each bit in each pattern is independently chosen to be either active
or inactive with equal probability and the same is assumed for the output
bits.

It has been known for some time (Cover, 1965) that such a set of random
associations is *typically* linearly separable, as long as the number P of patterns does not exceed twice the dimension N of the input space, $P \leq 2N$.
It turns out that the suitable representation of the active and inactive states
for this problem – i.e., appropriate for the given pattern statistics – is a ± 1
representation. Moreover, due to the symmetry between active and inactive
states in the problem, a zero threshold should be chosen.

Learning à la Hebb by correlating pre- and postsynaptic activities, one
has $(\Delta W_{0j})^\mu \propto \zeta_0^\mu \xi_j^\mu$ as the synaptic change in response to a presentation
of pattern μ. As we have mentioned already, this involves a modification of
Hebb's original proposal. Summing contributions from all patterns of the
problem set, one obtains [compare with (3)]

$$W_{0j} = \frac{1}{N} \sum_{\mu=1}^{P} \zeta_0^\mu \xi_j^\mu , \qquad (7)$$

where the prefactor is chosen just to fix scales in a manner that allows taking a sensible large system limit. Here we distinguish input- from output-bits by using different symbols for them. In recursive networks, outputs of single neurons are used as inputs by other neurons of the same net, and the distinction will be dropped in such a context.

It is not difficult to demonstrate that Hebbian learning finds an *approximation* to the separating hyperplane, which is rather good for small problem size P, but which becomes progressively worse as the number of patterns to be classified increases. To wit, taking an arbitrary example ξ^ν out of the set of learnt patterns, one finds that the Hebbian synapses (7) produce a local field of the form $h^\nu = h_0(\xi^\nu) = \zeta_0^\nu + \delta^\nu$. Here ζ_0^ν is the correct output-bit corresponding to the input pattern ξ^ν (the *signal*), which is produced by the ν-th contribution $(\Delta W_{0j})^\nu$ to the W_{0j}. The other contributions to h^ν do not add up constructively. Together they produce the *noise* term δ^ν. In the large system limit, one can appeal to the central limit theorem to show that the probability density of the noise is Gaussian with zero mean and variance $\alpha = P/N$.[2] A misclassification occurs, if the noise succeeds in reversing the sign determined by the signal ζ_0^ν. Its probability depends therefore only on α, the ratio of problem size P and system size N. It is exponentially small – $P_{err}(\alpha) \sim \exp(-1/2\alpha)$ – for small α, but increases to sizeable values already way below $\alpha_c = 2$, which is the largest value for which the problem is linearly separable, i.e., the largest value for which we know that a solution with $P_{err} = 0$ typically exists. If, however, a finite fraction of errors is tolerable, and such can be the case when one is interested in the overall output of a large array of perceptrons, then moderate levels of loading can, of course, be accepted. We shall see in Sect. 4 below that this is a standard situation in recursive networks.

The argument just presented can be extended to show that even distorted versions of the learnt patterns are classified correctly with a reasonably small error probability, provided the distortions are not too severe and, again, the loading level α is not too high.

The modified Hebbian learning prescription may be generalized to handle low activity data, i.e., patterns with unequal proportions of active and inactive bits. The appropriate learning rule is most succinctly formulated in terms of a 1-0 representation for the active and inactive states and reads

$$(\Delta W_{0j})^\mu \propto \tilde{\zeta}_0^\mu \tilde{\xi}_j^\mu , \qquad (8)$$

where $\tilde{\zeta}_0^\mu = \zeta_0^\mu - a_{out}$ and $\tilde{\xi}_j^\mu = \xi_j^\mu - a_{in}$, with $a_{in/out}$ denoting the probability of having active bits at the input and output sides, respectively. Non-zero

---

[2] The precise value is actually $(P-1)/N$.

thresholds are generally needed to achieve the desired linear separation. Interestingly this rule "approaches" Hebb's original prescription in the low activity limit $a_{\text{in/out}} \to 0$; the strongest synaptic changes occur if both presynaptic and postsynaptic neuron are active, and learning generates predominantly excitatory synapses. Interestingly too, this rule benefits from low activity at the output side: The variance of the noise contribution to local fields is reduced by a factor $a_{\text{out}}(1-a_{\text{out}})/(1-a_{\text{in}})$ relative to the case $a_{\text{in}} = a_{\text{out}} = \frac{1}{2}$, leading to reduced error rates and correspondingly enlarged storage capacities. We shall return to this issue in Sect. 4.

Two tiny modifications of the Hebbian learning rule (7,8) serve to boost its power considerably. First, synapses are only changed in response to a pattern presentation if the pattern is currently misclassified. If ζ_0^μ is the desired output bit corresponding to an input pattern $\boldsymbol{\xi}^\mu$ which is currently misclassified, then

$$(\Delta W_{0j})^\mu \propto \varepsilon^\mu \zeta_0^\mu \xi_j^\mu \,, \tag{9}$$

where ε^μ is an error mask that signifies whether the pattern in question is currently misclassified ($\varepsilon^\mu = 1$) or not ($\varepsilon^\mu = 0$). Here, a ± 1 representation for the output bits is assumed; the input patterns can be chosen arbitrarily in $\mathbb{R}^N$. Second, pattern presentation and (conditional) updating of synapses according to (9) is continued as long as errors in the pattern set occur. The resulting learning algorithm is called *percepton learning*.

An alternative way of phrasing (9) uses the output error $\delta_0^\mu = \zeta_0^\mu - \nu_0^\mu$, i.e., the difference between the desired ζ_0^μ and the current actual output bit ν_0^μ for pattern μ. This gives $(\Delta W_{0j})^\mu \propto \delta_0^\mu \xi_j^\mu$. It may be read as a combined process of learning the desired association and "unlearning" the current erroneous one.

Perceptron learning shares with Hebbian learning the feature that synaptic changes are determined by data locally available to the synapse – the values of input and (desired) output bits. Both the *locality* and the *simplicity* of the essentially Hebbian correlation-type synaptic updating rule must be regarded prerequisites if perceptron learning – indeed *any* learning rule – is to be considered as a "reasonable abstraction" of a biological learning mechanism.

Unlike Hebbian learning proper, perceptron learning requires a supervisor or teacher to compare current and desired performance. Here – as with any other supervised learning algorithm – we may encounter a problem, because neither do our synapses know about our higher goals, nor do we have immediate or deliberate control over our synaptic weights. It is conceivable though that the necessary supervision and feedback can be provided by other neural modules, provided that the output of the perceptron in question is "directly visible" to them and a more or less direct neural pathway for feedback is available. We will have occasion to return to this issue later on.

The resulting advantage of supervised perceptron learning over simple Hebbian learning, however, is dramatic. Perceptron learning is *guaranteed* to find a solution to a learning task after *finitely* many updatings, provided only that a solution exists, and no assumptions concerning pattern statistics need be made. Morevoer, learning of thresholds can, if necessary, be easily incorporated in the algorithm. This is the content of the so-called perceptron convergence theorem (Rosenblatt, 1962). For a precise formulation and for proofs, see Rosenblatt (1962), Minsky and Papert (1969), Hertz et al. (1991).

So far, we have discussed the problem of storing, or embedding a set of (random) associations in a perceptron. It is expedient to distinguish this problem from that of *learning a rule*, given only a set of examples representative of the rule.

For the problem of learning a rule, a new issue may be defined and studied, namely that of *generalization*. Generalization, as opposed to memorization, is the ability of a learner to perform correctly with respect to the rule in situations (s)he has not encountered before, i.e., during training.

For the perceptron, this issue may be formalized as follows. One assumes that a rule is given in terms of some unknown but fixed separating hyperplane according to which all inputs are to be classified. A set of P examples,

$$\zeta_0^\mu = \text{sgn}(\mathbf{W}^t \cdot \boldsymbol{\xi}^\mu) , \quad \mu = 1, \ldots, P , \tag{10}$$

is produced by a "teacher perceptron", characterized by its coupling vector $\mathbf{W}^t = (W_1^t, W_2^t, \ldots, W_N^t)$ which represents the separating hyperplane (the rule) to be learnt. That is, as before, the input patterns $\boldsymbol{\xi}^\mu$ are randomly generated; however, the corresponding outputs are now no longer independently chosen at random, but fixed functions of the inputs. A "student perceptron" attempts to learn this set of examples – called the *training set* – according to some learning algorithm.

The generalization error ε_g is the probability that student and teacher disagree about the output corresponding to a randomly chosen input that was not part of the training set. For perceptrons there is a very simple geometrical visualization for the probability of disagreement between teacher $\mathbf{W}^t$ and student $\mathbf{W}$. It is just $\varepsilon_g = \theta/\pi$, where θ is the angle between the teacher's and the student's coupling vector (see Fig. 4).

Assume that the student learns the examples according to the generalized Hebb rule. In vector notation,

$$\mathbf{W} = \frac{1}{N} \sum_{\mu=1}^{P} \zeta_0^\mu \boldsymbol{\xi}^\mu . \tag{11}$$

An argument in the spirit of the signal-to-noise-ratio analysis used above to analyse Hebbian learning of random associations can be utilized to obtain the generalization error as a function of the size P of the training set. To this end, one decomposes each input pattern $\boldsymbol{\xi}^\mu$ into its contribution parallel and orthogonal to $\mathbf{W}^t$. Through (11), this decomposition induces a corresponding decomposition of the student's coupling vector, $\mathbf{W} = \mathbf{W}_\parallel + \mathbf{W}_\perp$. Using

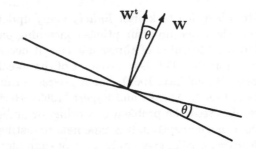

Fig. 4. Geometrical view of generalization

(10), one can conclude that the contributions to $\mathbf{W}_\parallel$ add up constructively, hence $\|\mathbf{W}_\parallel\|$ grows like $\alpha = P/N$ with the size P of the training set. The orthogonal contribution $\mathbf{W}_\perp$ to the student's coupling vector, on the other hand, can be interpreted as the result of an unbiased P-step random walk (a diffusion process) in the $N-1$-dimensional space orthogonal to $\mathbf{W}^t$, each step of length $1/\sqrt{N}$. So typically $\|\mathbf{W}_\perp\| \sim \sqrt{\alpha}$. In the large system limit, prefactors may be obtained by appeal to the central limit theorem, and the average generalization error is thereby found to be

$$\varepsilon_g(\alpha) = \frac{1}{\pi} \arctan(\|\mathbf{W}_\perp\|/\|\mathbf{W}_\parallel\|) = \frac{1}{\pi} \arctan\left(\sqrt{\frac{\pi}{2\alpha}}\right) . \qquad (12)$$

It decreases from $\varepsilon_g \simeq 0.5$ at the beginning of the training session – the result one would expect for random guesses – to zero, as $\alpha \to \infty$. The asymptotic decrease is $\varepsilon_g(\alpha) \sim 1/\sqrt{2\pi\alpha}$ for large α.

The simple Hebbian learning algorithm is thus able to find the rule asymptotically, although it is never perfect on the training set. A similar argument to that given for the generalization error can be invoked to compute the average training error ε_t, which is always bounded from above by the generalization error.

How does the perceptron algorithm perform on the problem of learning a rule. First, since the examples themselves are generated by a perceptron, hence linearly separable, perceptron learning is always perfect on the training set. That is $\varepsilon_t = 0$ for perceptron learning. To compute the generalization error, is not so easy as for the Hebbian student. We shall try to convey the spirit of such calculations later on in Sect. 3.3. Let us here just quote results.

Asymptotically the generalization error for perceptron learning decreases with the size of the training set as $\varepsilon_g(\alpha) \sim \alpha^{-1}$ for large α. The prefactor depends on further details. Averaging over all perceptrons which do provide a correct classification of the training set, i.e., over the so-called version space, one obtains $\varepsilon_g^{av}(\alpha) \sim 0.62/\alpha$. For a student who always is forced to find the best separating hyperplane for the training set (its orientation is such that the distance of the classified input vectors from either side of the plane is *maxi-*

mal) – this is the so-called optimal perceptron – one has $\varepsilon_g^{opt}(\alpha) \sim 0.57/\alpha$. It is known that the Bayesian optimal classifier (optimal with respect to generalization rather than training) has $\varepsilon_g^{Bayes}(\alpha) \sim 0.44/\alpha$, but this classifier itself is *not* implementable through a simple perceptron. Extensive discussions of these and related matters can be found in the literature (György and Tishby, 1990; Watkin et al., 1993; Opper and Kinzel, 1996; Engel, 1994).

Thus, perceptron learning generalizes faster than Hebbian learning, however at higher 'computational cost': the perceptron learner always has to retrain on the whole new training set every time a pattern is added to it. A significant computational effort is required on top of this if one always tries to find the optimal perceptron.

3.2 Layered Networks

To overcome the limitations of simple perceptrons so as to realize input–output relations more complicated than the linearly separable ones, one may resort to combining several simple perceptrons to build up more complicated architectures. An important class are the multi-layer networks to which we now turn.

In multi-layer networks, the output produced by a single perceptron is not necessarily communicated to the outside world. Rather one imagines a setup where several perceptrons are arranged in a layered structure, each node in each layer independently processing information according to its afferent synaptic weights and its transfer function Φ. The first layer – the input layer – receives input from external sources, processes it, and relays the processed information further through possibly several intermediate so-called hidden layers. A final layer – the output layer – performs a last processing step and transmits the result of the "neural computation" performed in the layered architecture to the outside world. Synaptic connections are such that no feedback loops exist.

Multi-layer networks consisting of *simple* perceptrons, each implementing a linearly separable threshold decision, have been discussed already' in the early sixties under the name of Gamba perceptrons (Minsky and Papert, 1969). For them, no general learning algorithm exists. The situation is different, and simpler, in the case where the elementary perceptrons making up the layered structure have a smooth, differentiable input–output relation. For such networks a general-purpose learning algorithm exists, which is guaranteed to converge at least locally to a solution, provided that a solution exists for the information processing task and the network in question.

The algorithm is based on gradient-descent in an "error–energy landscape". Given the information processing task – a set of input–output pairs $(\zeta_0^\mu, \boldsymbol{\xi}^\mu)$, $\mu = 1, \ldots, P$ to be embedded in the net – and assuming for simplicity a single output unit[3], one computes a network error measure over the set

---

[3] This implies no loss of generality. The problem may be analyzed separately for the sub-nets feeding each each output node.

of patterns

$$E = \frac{1}{2} \sum_\mu (\delta_0^\mu)^2 \ , \tag{13}$$

the output errors δ_0^μ being defined as before. For fixed input–output relations Φ, the error measure is determined by the set of all weights of the network $E = E(\mathbf{W})$. Let W_{ij} be a weight connecting node j to i. Gradient-descent learning aims at reducing E by adapting the weights W_{ij} according to

$$\Delta W_{ij} = -\eta \frac{\partial E(\mathbf{W})}{\partial W_{ij}} \ , \tag{14}$$

where η is a learning rate that must be chosen sufficiently small to ensure convergence to (local) minima of $E(\mathbf{W})$. For a network consisting of a single node, one has $\delta_0^\mu = \zeta_0^\mu - \nu_0^\mu$ with $\nu_0^\mu = \Phi(\sum_j W_{0j} \xi_j^\mu) = \Phi(h_0^\mu)$, hence

$$\Delta W_{0j} = \eta \sum_\mu \delta_0^\mu \Phi'(h_0^\mu) \xi_j^\mu \equiv \eta \sum_\mu \tilde{\delta}_0^\mu \xi_j^\mu \ , \tag{15}$$

where Φ' denotes the derivative of Φ. Note that there is a certain similarity with perceptron learning. The change of W_{0j} is related to the product of a (renormalized) error $\tilde{\delta}_0^\mu$ at the output side of node 0 with the input information ξ_j^μ, summed over all patterns μ.

If the network architecture is such that no feedback loops exist, this rule is immediately generalized to the multi-layer situation, using the chain rule of differential calculus. The resulting algorithm is called the *back-propagation* algorithm for reasons to become clear shortly. Namely, for an arbitrary coupling W_{ij} in the net one obtains

$$\Delta W_{ij} = -\eta \sum_\mu \tilde{\delta}_i^\mu \xi_j^\mu \ , \tag{16}$$

where ξ_j^μ is the input to node i in pattern μ, coming from node j (except when j denotes an external input line, this is not an input from the outside world), and $\tilde{\delta}_i^\mu$ is a renormalized output error at node i, computed by *back-propagating* the output errors of all nodes k to which node i relays *its* output via W_{ki},

$$\tilde{\delta}_i^\mu = \sum_k \tilde{\delta}_k^\mu W_{ki} \Phi'(h_i^\mu) \ . \tag{17}$$

Note that the (renormalized) error is propagated via the link $i \rightarrow k$ by utilizing that link in the *reverse* direction! This kind of error back-propagation needed for the updating of all links not directly connected to the output node is clearly biologically implausible. There is currently no evidence for mechanisms that might provide such functionality in real neural tissue.

Moreover, the algorithm always searches for the nearest local minimum in the error–energy landscape over the space of couplings, which might be a spurious minimum with an intolerably large error measure, and it would be stuck there. This kind of malfunctioning of the learning algorithm can to some extent be avoided by introducing stochastic elements to the dynamics which permit occasional uphill moves. One such mechanism would be provided by "online learning", in which the error measure is not considered as a sum of (squared) errors over the full pattern set, but rather as the contribution of the pattern currently presented to the net, and by training on the patterns in some random order.

Back-propagation is a very versatile algorithm, and it is currently the 'workhorse' for training multi-layer networks in practical or technical applications. The list of real-world problems, where neural networks have been successfully put to work, is already impressive; see, e.g., Hertz et al. (1991). Let us just mention two examples. One of the early successes was to train neural networks to read (and pronounce) written English text. One of the harder problems, where neural solutions have recently been found competitive or superior to heuristic engineering solutions, is the prediction of secondary structure of proteins from their amino-acid sequence. Both examples share the feature that algorithmic solutions to these problems are not known, or at least extremely hard to formulate explicitly. In these, as in many other practical problems, networks were found to generalize well in situations which were not part of the training set.

A generally unsolved problem in this context is that of choosing the correct architecture in terms of numbers of layers and numbers of nodes per layer necessary to solve a given task. Beyond the fact that a two-layer architecture is sufficient to implement continuous maps between the input- and output-side, whereas a three-layer net is necessary, if the map to be realized has discontinuities, almost nothing is known (Hertz et al., 1991). One has to rely on trial-and-error schemes along the rule of thumb that networks should be *as large as necessary, but as small as possible*, the first part addressing the representability issue, the second the problem that a neural architecture that is too rich will not be forced to extract rules from a training set but simply memorize the training examples, and so will generalize poorly. Algorithmic means to honour this rule of thumb in one way or another – under the categories of network-pruning or network-construction algorithms – do, however, exist (Hertz et al., 1991).

The situation is again somewhat better for certain simplified setups – two-layer Gamba perceptrons where the weights between a hidden layer and the output node are fixed in advance such that the output node computes a preassigned boolean function of the outputs of the hidden layer. Popular examples are the so-called committee machine (the output follows the majority of the hidden layer ouputs) and the parity machine (it produces the product of the ± 1 hidden layer outputs). For such machines, storage capacities and generalization curves for random (input) data have been computed, and the relevant scales have been identified: The number of random associations that can be embedded in the net is proportional to the number $\mathcal{N}$ of adjustable weights, and in order to achieve generalization, the size P of the training set must also be proportional to $\mathcal{N}$. The computations are rather involved and approximations have to be made, which are not in all cases completely under control. Moreover, checking by numerical simulations is hampered by the absence of good learning algorithms. So, whereas scales have been identified, prefactors are in some cases still under debate. A recent review is given by Opper and Kinzel (1996).

Neither back-propagation learning (online or offline) for general multi-layer networks nor existing proposals for learning in simplified multi-layer architectures of the kind just described [see, for instance, the review by Watkin et al. (1993)] can claim a substantial degree of biological plausibility. In this context it is perhaps worth mentioning a proposal of Bethge et al. (1994), who use the idea of fixing one layer of connections the other way round, and consider two-layer architectures with fixed input–to–hidden layer connections. These provide a preprocessing scheme which recodes the input data, e.g., by representing them locally in terms of mutually exclusive features. This requires, in general, a large hidden layer and divergent pathways. The advantage in terms of biological modeling, however, is twofold. There is some evidence that fixed preprocessing of sensory data which provides feature detection via divergent neural pathways is found in nature, for instance in early vision. Moreover, for learning in the second layer, simple perceptron learning can suffice, which – as we have argued above – still has some degree of biological plausibility to it. Quantitative analysis reveals that such a setup, one might call it coding machine, can realize mappings outside the linearly separable class (Bethge et al., 1994). The generalization ability of networks of this type remain to be analyzed quantitatively. It is clear, though, that the proper scale is again set by the number of adjustable units.

Interestingly, there exist unsupervised learning mechanisms that can provide the sort of feature extraction required in the approach of Bethge et al. Prominent proposals, which are sufficiently close to biological realism, are due to Linsker (1986) and Kohonen (1982; 1989). Linsker suggests a multilayer architecture of linear units trained via a modified Hebbian learning rule, for which he demonstrates the spontaneous emergence of synaptic connectivities that create orientation selective cells and so-called center-surround cells in upper layers, as they are also observed in the early stages of vision. Kohonen discusses two-layer architectures where neurons in the second layer "compete" for inputs coming from the first, which might be a retina. Lateral inhibition, i.e., feedback in the second layer ensures that only a single neuron in the second layer is active at a time, namely the one with the largest postsynaptic potential for the given input. An unsupervised adaptation process of synaptic weights connecting the input layer to the second layer is found to generate a system where each neuron in the second layer becomes active for a certain group of mutually similar inputs (stimuli). Note that this presupposes that similarity of, or correlations between different inputs exists. Inputs which are mutually similar, but to a smaller degree, excite nearby cells in the second layer. That is, one has feature extraction which preserves *topology*. Moreover, the resolution of the feature map becomes spontaneously finer for regions of the stimulus space in which stimuli occur more frequently than in others. Details can be found in Hertz et al. (1991).

3.3 A General Theoretical Framework for Analyzing Learning and Generalization

Let us close the present section with a brief and necessarily very schematic outline of a general theoretical framework in terms of which the issues of learning and generalization may be systematically studied. Not because we like to indulge in formalism, but rather because the theoretical framework itself adds interesting perspectives to our way of thinking about neural networks in general, which, incidentally, extends much further than our mathematical abilities to actually work through the formalism in all detail for the vast majority of relevant cases. Key ideas of the approach presented below can be traced back to pioneering papers of Elizabeth Gardner (1987, 1988).

To set up the theoretical framework, it is useful to describe the learning process in terms of a training energy. Assume that the task put to a network is to embed a certain set P of input–output pairs (ζ^μ, ξ^μ), $\mu = 1, \ldots, P$, where the output vectors ζ^μ may be determined from the input vectors ξ^μ according to some rule, or independently chosen. The training energy may then be written as

$$E(\mathbf{W}|\{\zeta^\mu, \xi^\mu\}) = \sum_\mu \varepsilon(\mathbf{W}, \zeta^\mu, \xi^\mu) \,, \qquad (18)$$

with a single pattern output error $\varepsilon(\mathbf{W}, \zeta^\mu, \xi^\mu)$ that is a nonnegative measure of the deviation between the actual network output $\nu^\mu = \nu(\mathbf{W}, \xi^\mu)$ and the desired output ζ^μ. In the case of recursive networks, more specifically, in the case of learning fixed point attractors in recursive networks, there is of course no need to distinguish between input and output patterns.

Learning by gradient descent in an error-energy landscape – that is learning as an optimization process – has been discussed above in connection with the back-propagation algorithm for feedforward architectures, where the absence of feedback–loops allowed to obtain rather simple expressions for the derivatives of E with respect to the W_{ij}. It was noted already in that context that, in order to avoid getting stuck in local suboptimal energy valleys, one may supplement the gradient dynamics with a source of noise. This would lead to the Langevin dynamics

$$\Gamma^{-1} \frac{\mathrm{d}}{\mathrm{d}t} W_{ij} = -\frac{\partial}{\partial W_{ij}} E(\mathbf{W}|\{\zeta^\mu, \xi^\mu\}) + \eta_{ij}(t) \,, \qquad (19)$$

in which the (systematic) drift term aims to reduce the training error, whereas the noise allows occasional changes for the worse.

There is more to adding noise than its beneficial role in avoiding suboptimal solutions. Namely, if the noise in (19) is taken to be uncorrelated Gaussian white noise, with average $\langle \eta_{ij}(t) \rangle = 0$ and covariance $\langle \eta_{ij}(t)\eta_{kl}(t') \rangle = 2T\delta_{(ij),(kl)}\delta(t - t')$, then the Langevin dynamics (19) is known to converge asymptotically to 'thermodynamic equilibrium' described by a Gibbs distribution over the space of synaptic weights,

$$\mathcal{P}(\mathbf{W}|\{\boldsymbol{\zeta}^{\mu},\boldsymbol{\xi}^{\mu}\}) = \mathcal{Z}^{-1}\exp\{-\beta E(\mathbf{W}|\{\boldsymbol{\zeta}^{\mu},\boldsymbol{\xi}^{\mu}\})\} . \tag{20}$$

Here β denotes an inverse temperature[4] in units of Boltzmann's constant, $\beta = 1/T$. In the case where the W_{ij} are only allowed to take on discrete values, the Langevin dynamics (19) would have to be replaced by a Monte-Carlo dynamics at finite temperature, the analog of gradient descent being realized in the limit $T \to 0$. The equilibrium distribution would still be given by (20), provided the transition probabilities of the discrete stochastic dynamics were properly chosen. Note that $\mathcal{P}$ depends parametrically on the choice of training examples.

Now two interesting things have happened. First, by introducing a suitable form of noise and by considering the long time limit of the ensuing stochastic dynamics, we know the distribution $\mathcal{P}$ over the space of weights explicitly, so we can in principle compute averages and fluctuations of *all* observables for which we know how they depend on the W_{ij}. Second, by considering the equilibrium distribution (20), one is looking at an "ensemble of learners" which have reached, e.g., a certain average asymptotic training error, and one is thereby *deemphasizing all details of the learning mechanism that may have been put to work to achieve that state.* This last circumstance is one of the important sources by which the general framework acquires its predictive power, because it is more likely than not that we do not know the actual mechanisms at work during learning, and so it is gratifying to see that at least asymptotically the theory does not require such knowledge.

Of the quantities we are interested in computing, one is the average training error

$$\langle E \rangle = \langle E(\mathbf{W}|\{\boldsymbol{\zeta}^{\mu},\boldsymbol{\xi}^{\mu}\}) \rangle = \int d\mu(\mathbf{W}) \, \mathcal{P}(\mathbf{W}|\{\boldsymbol{\zeta}^{\mu},\boldsymbol{\xi}^{\mu}\}) \, E(\mathbf{W}|\{\boldsymbol{\zeta}^{\mu},\boldsymbol{\xi}^{\mu}\}) , \tag{21}$$

where the measure $d\mu(\mathbf{W})$ encodes whatever a priori constraints might be known to hold about the W_{ij}. It may also be obtained from the "free energy"

$$\mathcal{F} = -\beta^{-1}\ln \mathcal{Z} = -\beta^{-1}\ln \int d\mu(\mathbf{W}) \exp\{-\beta E(\mathbf{W}|\{\boldsymbol{\zeta}^{\mu},\boldsymbol{\xi}^{\mu}\})\} \tag{22}$$

corresponding to the Gibbs distribution (20) via the thermodynamic relation

---

[4] Note that we use temperature T not as specifying ambient temperature, but simply as a measure of the degree of stochasticity in the dynamics.

$$\langle E \rangle = \frac{\partial}{\partial \beta} \beta \mathcal{F} \; . \tag{23}$$

The result still depends on the (random) examples chosen for the training set, so an extra average over the different possible realizations of the training set must be performed, which gives

$$E(\beta, P) = \langle\!\langle\langle E \rangle\rangle\!\rangle = \int \prod_{\mu} \mathrm{d}\rho(\zeta^{\mu}, \xi^{\mu}) \, \langle E \rangle \; . \tag{24}$$

Such an average is automatically implied if one replaces the free energy in the thermodynamic relation (23) by its average over the possible training sets, i.e., the so-called quenched free energy $\mathcal{F}_q = -\beta^{-1} \langle\!\langle \ln \mathcal{Z} \rangle\!\rangle$. Similarly, the average generalization error is obtained by first considering $\varepsilon_g(\mathbf{W}) = \int \mathrm{d}\rho(\zeta, \xi) \, \varepsilon(\mathbf{W}, \zeta, \xi)$, that is, the single pattern output error used in (19), averaged over all possible input output pairs which were not part of the training set, and by computing

$$\varepsilon_g(\beta, P) = \langle\!\langle\langle \varepsilon_g(\mathbf{W}) \rangle\rangle\!\rangle \; . \tag{25}$$

Actually, it turns out that the additional averaging over the various realizations of the training set need not really be performed, because each training set will typically produce the same outcome, which is therefore called self-averaging. Technically, however, such averages are usually easier to handle than specific realizations, and the averages are therefore nevertheless computed. The same situation, incidentally, is encountered in the analysis of disordered condensed matter systems. Not too surprisingly therefore, it is this subdiscipline of physics from which many of the technical tools used in quantitative analyses of neural networks have been borrowed.

It is well known that the statistical analysis of conventional condensed matter comes up with virtually deterministic relations between macroscopic observables characteristic of the systems being investigated, as their size becomes large (think of relations between temperature, pressure, and density, i.e., equations of state for gases). In view of the appearance of relations of statistical thermodynamics in the above analysis, one may wonder whether analogous deterministic relations might emerge in the present context. This is indeed the case, and it may be regarded as the second source of predictive power of the general approach.

In the large system limit, that is, as the number $\mathcal{N}$ of synaptic couplings becomes large, the distribution (20) will give virtually all weight to $\mathbf{W}$-configurations with the same macroscopic properties. Among these are, in particular, the training error per pattern, $\varepsilon_t = P^{-1} \sum_{\mu} E(\mathbf{W}|\{\zeta^{\mu}, \xi^{\mu}\})$, and the generalization error ε_g.

The analysis reveals that a proper large system limit generally requires the size P of the training set to be scaled according to $P = \alpha \mathcal{N}$, as we have observed previously in specific examples. As $\mathcal{N} \to \infty$ (at fixed α) learning and generalization errors are *typically* – i.e., for the overwhelming majority of realizations – given by their thermodynamic averages (as functions on the α–scale), $\varepsilon_t = P^{-1}E(\beta, P) \to \varepsilon_t(\beta, \alpha)$ and $\varepsilon_g \to \varepsilon_g(\beta, \alpha)$.

The reason why the generalization error is among the predictable macroscopic quantities stems from the fact that it is related to the distance in weight space, $\Delta(\mathbf{W}^t, \mathbf{W}) = \mathcal{N}^{-1}\sum_{ij}(W_{ij}^t - W_{ij})^2$, between the network configuration $\mathbf{W}$ and the target configuration $\mathbf{W}^t$ which the learner is trying to approximate. This is itself a (normalized) extensive observable which typically acquires non-fluctuating values in the thermodynamic limit.

The results obtained via the statistical mechanics approach are, as we have indicated, *typical* in the sense that they are likely to be shared by the vast majority of realizations. This is to be seen in contrast to a set of results about learning and generalization, obtained within the machine-learning community under the paradigm of "probably almost correct learning". They usually refer to worst-case scenarios and do, indeed, usually turn out to be overly pessimistic. We refer to Watkin et al. (1993), Engel (1994), and Opper and Kinzel (1996) for more details on this matter.

In the zero-temperature ($\beta \to \infty$) limit, the Gibbs distribution (20) gives all weight to the synaptic configurations which realize the smallest conceivable training error. An interesting question to study in this context is what the largest value of α is, such that the minimum training energy is still zero. This then gives the size of largest pattern set that can be embedded without errors in the given architecture – *irrespective* of the learning algorithm used to train the net. This number is called the absolute capacity of the net, and it depends, of course, on the pattern statistics. In the case where outputs in the pattern set are generated according to some rule, one obtains information as to whether or not the rule is learnable, i.e., representable in the network under consideration.

For unbiased binary random patterns, the absolute capacity is found to be $\alpha_c = 2$ for networks consisting of simple threshold elements, and without hidden neurons. The number increases if the patterns to be embedded in the net have unequal proportions of active and inactive bits (see also Sect. 4 below); it decreases if one wants to embed patterns with a certain stability, that is, such that correct classifications are obtained even with a certain amount of distortion at the input side (Gardner, 1987, 1988). In attractor networks, high stability implies large basins of attraction for the patterns embedded in the net.

Another way to phrase these ideas is to note that learning of patterns puts restrictions on the allowed synaptic couplings. The absolute capacity is reached when the volume of allowed couplings, which becomes progressively smaller as more and more patterns are being embedded in the net, eventually shrinks to zero. The logarithm of the allowed volume is like an entropy, a measure of diversity. Learning then reduces the allowed diversity in the space of (perfect) learners. Similarly, by learning a rule from examples, the volume in the space of couplings will shrink with increasing size of the training set, and eventually be concentrated around the coupling vector representative of the target rule. Generalization ensues.

An interesting application of these ideas as means to predict the effects of brain lesions has been put forward by Virasoro (1988). He demonstrated that after learning hierarchically organized data – items grouped in classes of comparatively large similarity within classes, and greater dissimilarity between classes – the class information contained in each pattern enjoys a greater embedding stability than the information that identifies a pattern as a specific member of a class. As a consequence, brain lesions that randomly destroy or disturb a certain fraction of synapses after learning will lead to the effect that the *specific* information is lost first, and the class information only when damage becomes more severe. An example of the ensuing kind of malfunctioning is provided by the prosopagnosia syndrome – characterized by the ability of certain persons to recognize faces as faces, without being able to distinguish between individual faces. According to all we have said before, this kind of malfunctioning must *typically* be expected to occur in networks storing hierarchically organized data that have been injured. Note moreover that, beyond the fundamental supposition that memory resides in the synaptic organization of a net, hardly anything else has to be assumed for this analysis to apply.

It is perhaps worth pointing out that the Gibbs distribution (20) enjoys a distinguished status in the context of maximum-entropy/minimum-bias ideas (Jaynes, 1979). It is the maximally unbiased distribution of synaptic couplings, subject only to an, at least in principle, observable constraint, namely that of giving rise to a certain average training error. Together with the notion of concentration of probabilities at entropy maxima (Jaynes, 1979), this provides yet another source of predictive power that may be attributed to the general scheme.

Finally, we should not fail to notice that there is, of course, also room and need for studying learning dynamics proper as opposed to the statistics of asymptotic solutions, because information about final statistics tells us nothing about the time needed to reach asymptotia, which is also relevant and important information, certainly in technical applications. Here, we content ourselves with quoting just one pertinent example. The *existence* of neural solutions for a given storage task, which may be investigated by considering the allowed volume in the space of couplings, tells us nothing about our ability to find them. For the perceptron with binary weights, for instance, Horner (1992) has demonstrated that algorithms with a complexity scaling polynomially in system size are not likely to find solutions at *any* non-zero value of α in the large system limit, despite the fact that solutions are known to exist up to $\alpha_c \simeq 0.83$.

4 Attractor Networks – Associative Memory

Memory is one of the basic functions of our brain and it also plays a central role in any computing device. The memory in a computer is usually organized such that different contents are stored under different addresses. The address itself, typically a number, has no relation to the information which is found under its name. The retrieval of information requires the knowledge of the corresponding address or additional search engines using key words with lists of addresses and cross references.

An associative memory is a device which is organized such that part of the information allows recall of the full information stored. As an example the scent of a rose or the spoken word 'rose' recalls the full concept *rose*, typical forms and colors of its blossoms and leaves, or events in which a rose has played a role.

On a more abstract level we would like to have a device in which certain patterns ξ^μ are stored and where a certain input η recalls the pattern closest to it. This could be achieved by searching through the whole set of memories, but this would be rather inefficient.

A neural network is after all a dynamical system. Its dynamics could be defined by the update rule (4) or equivalently by a set of nonlinear differential equations

$$\frac{\mathrm{d}\nu_i(t)}{\mathrm{d}t} = -\frac{1}{\bar\tau}\left(\nu_i(t) - \Phi\left(\sum_j W_{ij}\nu_j(t) - \vartheta\right)\right) , \qquad (26)$$

where $\bar\tau$ is some average delay time. It is known from the theory of dynamical systems that equations of this type have attractors. That is, any solution with given initial values approaches some small subset of the full set of available states, which could be a stationary state (fixed point), a periodic solution (limit cycle) or a more complicated attractor. The set of initial values giving rise to solutions approaching the same attractor is called the basin of attraction of this attractor. This can now be used to construct an associative memory, if we succeed in finding synaptic couplings such that the patterns to be stored become attractors. If this is achieved, an initial state not too far from one of the patterns will evolve towards this pattern (attractor), provided it was within its basin of attraction.

It is clear that this mechanism requires networks with strong feedback. In a feedforward layered network with well defined input and output layers, the information would simply be passed from the input layer through hidden layers to the output layer, and without input such a network would be silent.

The goal is not only to find the appropriate couplings using a suitable learning rule, but also to estimate how many patterns can be stored and how wide the basins of attractions are. Wide basins of attraction are desirable because initial states having a small part in common with the pattern to be retrieved should be attracted by this pattern.

4.1   The Hopfield Model

A great deal of qualitative and quantitative understanding of such associative memories has come from a model proposed by Hopfield (Hopfield, 1982; Amit, 1989; Hertz et al., 1991). Its purpose is to store uncorrelated binary random patterns $\xi_i^\mu = \pm 1$, where $i = 1, \ldots, N$ labels the nodes (neurons) and $\mu = 1, \ldots, P$ the patterns to be stored. It employs the modified Hebb learning rule (3)

$$W_{ij} = \frac{1}{N} \sum_\mu \xi_i^\mu \xi_j^\mu , \qquad (27)$$

and one assumes that each node is connected with every other node. For the dynamics one uses a discretized version of (26), picking a node i at random and updating its value according to

$$\nu_i(t + \bar{\tau}) = \text{sgn}\Big(\sum_{j(\neq i)} W_{ij} \nu_j(t) \Big) . \qquad (28)$$

For the analysis of this model it is useful to define an 'energy' or 'cost function'

$$E(t) = -\frac{1}{2} \sum_{ij} \nu_i(t) \, W_{ij} \, \nu_j(t) \qquad (29)$$

for the firing pattern $\nu_i(t)$ at a given time t. It can easily be shown that this function can never increase in the course of time. This implies that the firing pattern will evolve in such a way that the system approaches one of the minima of E. This is like moving in a landscape with hills and valleys, and going downhill until a local minimum is reached. The existence of such a function, called Lyapunov function, ensures that the only attractors of such a model are fixed points or in the present context stationary firing patterns.

It has to be shown now that, with the above learning rule, the attractors are indeed the patterns to be stored, or at least close to them. The arguments are similar to those given in the context of the perceptron. As a measure of the distance between the actual state and a given pattern we introduce the 'overlap'

$$m_\mu(t) = \frac{1}{N} \sum_i \xi_i^\mu \nu_i(t) \qquad (30)$$

which is less than or equal to one, with $m_\lambda(t) = 1$ signifying that the actual firing pattern is that of pattern λ. If this is the case, the overlap with all the other patterns will be of order $1/\sqrt{N}$. Using the overlap, we can write the energy as

$$E(t) = -\frac{1}{2} \sum_\mu m_\mu(t)^2. \qquad (31)$$

In the limit of large N, and considering an initial state such that the initial overlap $m_\lambda(0)$ is the only one which is of order 1, the remaining ones being of order $1/\sqrt{N}$, one may approximate the energy by $E(t) \simeq -m_\lambda(t)^2/2$, assuming that $m_\lambda(t)$ remains the only finite overlap for all time. If this is the case, the energy will decrease and reach its minimal value for $m_\lambda(t) \to 1$, as $t \to \infty$. That is, the network has reconstructed pattern λ.

For initial states having a finite overlap with more than one pattern, the attractor reached can be a new state, called spurious state, composed of parts of several learnt patterns (Amit et al., 1985; Amit, 1989). This tells us that the network seems to memorize patterns which have not been learnt. It is not clear whether this has to be considered as malfunctioning or whether it gives room for creativity in the sense of novel combinations of acquired experience. With a slightly modified dynamics (Horner, 1987), a mixed initial state can also evolve towards the pattern with maximal initial overlap. Depending on the overall situation a network might switch from one mode to the other.

The picture so far presented holds as long as the loading $\alpha = P/N$ is small enough, so that the random contributions to the energy due to the $m_\mu(t) \sim 1/\sqrt{N}$ with $\mu \neq \lambda$ can be neglected.

For higher loading, the influence of these remaining patterns has to be taken into account. A more thorough investigation (Amit et al., 1985; Amit, 1989; Hertz et al., 1991) shows that this has two effects. First of all the retrieval states (minima of E) are no longer exactly the learnt patterns, but close to them with a small number of errors. For the whole range of loadings for which this kind of memory works, the final overlap is larger than 0.96, increasing with decreasing loading. In addition new attractors are created having a small or no overlap with any of the patterns. Their effect is primarily (Horner et al., 1989) to narrow the basins of attraction of the learnt patterns. At a critical loading of $\alpha_c \simeq 0.138$ these states cause a sudden breakdown of the whole memory.

This sudden breakdown due to overloading can be avoided by modified learning rules. Depending on details (see Chap. 3 of Hertz et al., 1991) either the earliest or the most recent memories are kept and the others are forgotten. It is also possible to keep the earliest and the most recent memories and to forget those in between, which seems to be the case with our own memory. Furthermore certain memories can be strengthened or erased by unconscious events taking place for instance during dream phases (see Chap. 3 of Hertz et al., 1991).

In order to estimate how efficiently such a memory works, it is not only necessary to find out how many patterns can be stored and how many errors the retrieval states have, it is also necessary to investigate the size of the basins of attraction, in other words, the fraction of a pattern that has to be offered as initial stimulus in order to retrieve this pattern. An investigation of the retrieval process (Horner et al., 1989) shows that this minimal initial overlap depends on the loading α, and for $\alpha < 0.1$ one finds the approximate retrieval condition $m_\lambda(0) > 0.4\,\alpha$. Finally, one can also estimate the gain of information reached during retrieval. This is the difference between the information contained in the pattern retrieved and the information that must be supplied in the initial stimulus to guarantee successful retrieval. This again depends on the loading, and a maximum of 0.1 bit per synapse is reached for $\alpha \approx 0.12$.

Another quantity of interest is the speed of retrieval. One finds that almost complete retrieval is already reached after only 3 updates per node. Inserting numbers for the relevant time scales of neurons one obtains 30–60 ms. This can be compared to measured reaction times which are typically of the order of 100–200 ms.

Among other reasons, the Hopfield model is unrealistic in the sense of requiring complete and symmetric connectivity. The requirement of symmetry $W_{ij} = W_{ji}$ ensures, in particular, the existence of an energy or cost function (29) ruling the dynamics of the network. The connectivity among cortical neurons is high, of the order of 10^4 synapses per neuron, but far from being complete, keeping in mind that within the range of the dendritic tree of a single neuron not more than approximately 10^5 other neurons are found. This has been taken into account in a study (Derrida et al., 1987) of a model with randomly diluted synaptic connections. The overall properties remain unchanged. The maximal number of patterns is now proportional to the average number C of afferent synapses per neuron, $P_{\max} = \alpha_c C$, with $\alpha_c \simeq 0.64$, but the total gain of information per synapse is still similar to the value obtained for the original model. A different behavior is found as the critical loading, α_c, is approached: In this model the basins of attraction remain wide, but the number of errors in the retrieval state increases drastically as $\alpha \to \alpha_c$.

4.2 Sparse Coding Networks

As mentioned previously, a remarkable feature of cortical neurons is their low average firing rate. In principle a neuron can produce as many as 300 spikes per second. Recordings on living vertebrate's brains typically show some cells firing at an elevated rate of up to 30 spikes per second, but the average rate is much lower, only 1 to 5 spikes per second. Retaining the proposal of rate coding one has to conclude that typical firing patterns are sparse in the sense that the number of active neurons $N_a(t)$ at each time is much smaller than the number of silent neurons $N_s(t)$. This means that the mean activity $a(t) = N_a(t)/(N_a(t) + N_s(t))$ is low.

Various versions of attractor networks with low activity have been investigated (Willshaw et al, 1969; Palm, 1982; Tsodyks and Feigel'man, 1988; Amit et al., 1994, 1997) in the literature. Within the framework of binary McCulloch–Pitts neurons their state is conveniently represented by $\nu_a = 1$ for active and $\nu_s = 0$ for silent neurons. In this case the original Hebbian learning rule (2,3) reinforcing the coupling strength between neurons active at the same time is appropriate.

Obviously this learning rule creates only excitatory synaptic connections, so in addition inhibitory neurons are required to control the mean activity of the network, as discussed in Sect. 2. It turns out that this control has to be faster than the action of the excitatory synapses. This seems to be supported by the findings that the connections with inhibitory neurons are short and their synapses are typically attached to the soma or the innermost parts of the dendrites of the excitatory pyramidal cells.

The update rule (26,28) has to be modified according to the $(0;1)$ representation using a step function $\Phi(x) = 1$ for $x > 0$ and $\Phi(x) = 0$ otherwise.

Again such networks can serve as fast associative memories. The maximal loading depends on the mean activity. It diverges as $\alpha_c \sim 1/a \ln(1/a)$ for $a \to 0$. At the same time the information per pattern decreases with decreasing activity such that the total gain of information reaches a constant value of 0.72 bit per synapse (Horner et al., 1989). This value, however, is reached very slowly; for example, at $a = 0.001$ one finds $\alpha_c = 30$ and only 0.3 bit information gain per synapse. Nevertheless, this value exceeds the one found for the Hopfield model.

It should be noted that the class of low activity networks just described only solves the *spatial* aspect of the low activity issue. However, by going one step further and returning to the continuous-time dynamics (26), and by using more realistic 'graded' neural input–output relations, one can solve the temporal aspect as well. Neurons which should be firing in one of the low activity attractors are then typically found to fire also at low rate (Kühn and Bös, 1993). Within models of neural networks based on spiking neurons, this issue has been addressed by Amit et al. (1994, 1997).

One of the virtues of sparse coding networks is in the learning rule. A change in the synaptic strength is required only if both, the pre- and the postsynaptic neuron are active at the same time. This implies that the total number of learning events is reduced compared to a network with symmetric coding and consequently the requirements on accuracy and reproducibility of each individual learning process are less stringent.

Another reason why nature has chosen sparse coding could of course be reduction of energy consumption because each spike requires some extra energy beyond the energy necessary to keep a neuron alive.

In a sparse coding network it makes sense to talk about the foreground of a pattern, made up of the active neurons in this pattern, and a background containing the rest. The foreground is usually denoted as cell assembly, a notion which goes back to Hebb. The probability that a neuron belongs to the foreground of a given pattern is given by the mean activity a of this pattern, which is assumed to be low. The probability that this neuron belongs simultaneously to the foreground of two patterns is given by a^2. This means that the cell assemblies belonging to different patterns are almost completely disjoint. As a consequence mixture states are no problem because their mean activity is higher and they can be suppressed by the action of the inhibitory neurons regulating the overall activity. This will play a role for some of the functions discussed later.

4.3   Dynamical Attractors

The attractor network models discussed so far allowed only for fixed point attractors or stationary patterns as retrieval states. This is a severe restriction and one can think of many instances where genuine dynamical attractors are called for. The reason for the restriction is the existence of a Lyapunov function which can be traced back to the symmetry of the couplings $W_{ij} = W_{ji}$. This shows that asymmetric couplings have to be included if dynamical attractors are to be constructed (see Chap. 3 of Hertz et al., 1991).

Let us demonstrate this again on a somewhat artificial example. The desired attractor should be composed of a sequence of patterns ξ_i^μ such that pattern μ is present for some time τ and then the next pattern $\mu + 1$ is presented. The whole set of patterns with $\mu = 1, \ldots, L$ can be closed such that pattern 1 is shown again after the last pattern L has appeared, generating a periodically repeating sequence. This is called a limit cycle. The retrieval of this cycle should work such that the network is initialized by a firing pattern close to one of the members of the cycle, say pattern 1, this pattern is completed, and after a time τ pattern 2 appears and so on.

This can be achieved by using two types of synapses, fast synapses W_{ij}^f without delay and slow synapses W_{ij}^s with delay τ. The update rule (26) now reads

$$\frac{d\nu_i(t)}{dt} = -\frac{1}{\tau}\left(\nu_i(t) - \Phi\Big(\sum_j W_{ij}^f \nu_j(t) + \sum_j W_{ij}^s \nu_j(t - \tau) - \vartheta\Big)\right). \quad (32)$$

The appropriate choice of the couplings is [see Eq. (27)]

$$W_{ij}^f = \frac{1}{N}\sum_{\mu=1}^{L} \xi_i^\mu \xi_j^\mu \quad \text{and} \quad W_{ij}^s = \frac{\lambda}{N}\sum_{\mu=1}^{L} \xi_i^{\mu+1} \xi_j^\mu \quad (33)$$

with pattern $L + 1$ being equivalent to pattern 1.

Assume that the network was in a random state for $t < 0$ and has been brought into a state close to pattern 1 at $t = 0$. For $0 < t < \tau$ the slow asymmetric synapses will have no effect, whereas the fast synapses drive the state even closer to pattern 1. For $\tau < t < 2\tau$ the slow synapses now tend to drive the state from pattern 1 to pattern 2, and if they are stronger than the fast synapses ($\lambda > 1$), the state actually switches to pattern 2, which is then reinforced by the action of the fast synapses as well. This process is repeated and the whole cycle is generated.

Obviously due to the cyclic symmetry any pattern of the cycle can be used for retrieval. Furthermore it is possible to store more than one cycle or cycles and fixed points in the same network. For the storage capacity the total number of patterns in all attractors is crucial.

The decisive step in this model is the addition of the non-symmetric slow synapses which ultimately cause the switching between successive patterns. Devices of this kind have been studied in several variations (see Chap. 3 of Hertz et al., 1991).

The mechanism sketched above requires the existence of slow synapses having exactly the delay time necessary for the desired timing of the attractor. This can easily be relaxed (Herz et al., 1989) by assuming a pool of synapses W_{ij}^τ with different delays τ. Employing a modified Hebb learning rule (2)

$$\Delta W_{ij}^\tau(t) \propto \nu_i(t)\,\nu_j(t-\tau)\;,\tag{34}$$

the training process reinforces specifically those synapses which have the appropriate delay time and cycles with different times for the presentation of each individual pattern can be learnt. This learning rule is actually the natural extension of Hebb's idea, assuming that the delay is caused primarily by the axonal transmission time.

One can think of other mechanisms to determine the speed at which consecutive patterns are retrieved. One such mechanism (Horn and Usher, 1989) uses the phenomenon of fatigue or adaptation (see Sect. 2) and some special properties of sparse coding networks. The process of adaptation can be mimicked by a time dependent threshold $\vartheta_i(t)$ with

$$\frac{\mathrm{d}\vartheta_i(t)}{\mathrm{d}t} = \frac{1}{\tau_a}\Big(\vartheta_0 + \vartheta'\nu_i(t) - \vartheta_i(t)\Big)\tag{35}$$

where τ_a is the time constant relevant for adaptation. According to this equation the threshold of a silent neuron relaxes towards ϑ_0 and is increased if this neuron fires at some finite rate.

For the synaptic couplings again a combination of symmetric couplings, stabilizing the individual patterns, and non-symmetric couplings, favoring transitions to the consecutive patterns in the sequence, is used. This means that eqs.(32,33) can again be used with the above time dependent threshold $\vartheta_i(t)$ but without retardation in the asymmetric couplings W_{ij}^s. In contrast to the above model, $\lambda < 1$ has now to be chosen.

This works as follows. Assume the network was in a completely silent state for $t < 0$ and all the thresholds have their resting value ϑ_0. Applying an external stimulus exciting the cell assembly or pool of active neurons of pattern 1, the symmetric couplings stabilize this pattern. The nodes which should be active in pattern 2 are also excited but if λ is sufficiently small the action of the asymmetric couplings is not strong enough to make them fire, too. As time goes on, the neurons active in pattern 1 adapt and their threshold increases, reducing their firing rate. This also reduces the global inhibition and at some time the action of the weaker asymmetric couplings will be strong enough to activate the pool of neurons which have to be firing in pattern 2. This of course only works if the neurons of this second pool are still fresh. This is the case , however, because in a sparse coding network the probability of finding a neuron simultaneously in the cell assemblies of two consecutive patterns is low. After adaptation of the neurons in the second pool the state switches to pattern 3 and so on.

4.4   Segmentation and Binding

A similar sparse coding network with adaptive neurons can also solve the problem of segmentation (Horn and Usher, 1989; Ritz et al., 1994). Assume that an external stimulus simultaneously excites the pools of neurons of more than one pattern. The task is then to exhibit the separate identity of these patterns despite the fact that their representative neuron pools are simultaneously excited. This can be achieved by *activating*, i.e., retrieving only one of the patterns at a time and selecting another one a bit later. This is actually what we do when we are confronted with complex situations containing several unrelated objects. We concentrate on one object for some time and then go to the next, and so on.

A sparse coding network with suitable inhibition will allow for the activation of a single pattern only, because the simultaneous recall of two or more patterns would create an enhanced overall activity which is suppressed by the action of the inhibitory neurons. If exposed to a stimulus containing more than one learnt pattern, this network will first activate the pattern having the strongest input. After some time the pool of active neurons in this pattern will have adapted and the network retrieves the pattern with the second strongest stimulus because its pool of neurons is still fresh, disregarding again the small number of neurons common to the active pools of both patterns. This goes on until all patterns contained in the external stimulus have been retrieved or until the neurons of the first pool have recovered sufficiently to be excited again. Due to this recovery only a small number of patterns can be retrieved one after the other, and those being weakly stimulated may never appear. This is in accordance with our everyday experience.

This example of course is not a proof that this is how segmentation is done in our own brain. It only shows how it could plausibly be done. This critique, however, applies to the other models discussed as well.

A complementary problem is that of binding. Imagine a visual scene in which a large object moves behind some obstacle. What is actually seen is the front and the back end of this object, with the middle part hidden. The feature which is common to both parts is the speed at which they move, and this allows one to identify both parts as belonging to one object. If only one part is moving, they are easily identified as parts of two different objects. That is, parts of a complex stimulus having certain features in common are identified as parts of a larger object; these parts are linked.

A possible mechanism for this linking was discovered in multi-electrode recordings in the visual cortex of cats and other mammals (Gray and Singer,

1989; Eckhorn et al., 1988). It was observed that a moving light bar creates an oscillatory firing pattern in the cells having appropriate receptive fields. A second light bar created an oscillatory response in some other neurons. The motion of both bars in the same direction created a synchronization and phase locking of these oscillations whereas no such effect was observed if they were moved in different directions. This effect could even be observed among neurons belonging to different areas in the visual cortex.

The proposal is now that linking is performed by synchronization of oscillatory or more general firing patterns.

The observed oscillations had a period of about 20 ms and lasted for about 10 periods. Synchrony was already established within the first few oscillations. It should be pointed out that an individual neuron emits at most one or two spikes during one period. This means that larger assemblies' of cells with similar receptive fields have to cooperate. Actually the oscillations were observed in intercellular recordings which pick up the signals of many adjacent neurons, or in averages over many runs. The fast synchronization time and the relatively short duration of the oscillations might indicate that the important feature is not so much the existence of these oscillations, but rather the synchronous activity within a range of a few milliseconds.

Not too surprisingly several idealized models have been proposed reproducing this effect. Most of them are still based on a rate coding picture. This seems problematic in view of the short times involved and the relatively low average spiking rates of any individual neuron. Nevertheless rate coding is not completely ruled out if one keeps in mind that a rate has to be understood not as a temporal average over a single cell but rather as an average over assemblies of similar cells.

4.5   Synchronization of Spikes and Synfire Chains

Rate coding is the widely accepted paradigm for the predominant part of data processing in the brain of vertebrates. Keeping in mind that rates might have to be understood as averages over groups of neurons, elementary operations could be performed within the integration time of a neuron, typically 10 ms. The exact timing of the incoming spikes within this period should not matter.

If, on the other hand, a short volley of synchronized spikes arrives at a neuron within a fraction of a millisecond, this neuron can fire within a fraction of a millisecond. This can be used for very fast data processing whenever necessary, for instance in the auditory pathway where phase differences in the signals coming from the two ears are analyzed.

This raises the issue of whether such short volleys of synchronized spikes are a general feature, and what new kind of data processing can be achieved in this way.

A possible such mechanism are synfire chains (Abeles, 1991). Their building blocks are pools of neurons locally connected in a feedforward manner. If the neurons in one pool are stimulated simultaneously, they will emit synchronized spikes. After some short delay time these spikes arrive at the neurons forming the next pool and cause a synchronous firing of this pool too. This process is repeated and a wave of activity travels with a certain speed along the chain. Actually the neurons forming the pools are all members of a larger network and a given neuron can belong to several pools. The chain and its pools are only defined by their connectivity. There might be also connections from the neurons of one pool to other neurons not belonging to the next pool. These connections have to be weak, however, otherwise those postsynaptic neurons have to be counted as members of the next pool. The picture of distinct pools is somewhat washed out if variations in the delay times are taken into account. What matters is the synchronous timing of the incoming spikes.

In some sense the idea of synfire chains is closely related to the dynamic attractors discussed earlier. The difference is in the sharp synchronization of the volleys of spikes. Model calculations show that some initial jitter in the volleys can even be reduced and synchrony sharpened up, stabilizing the propagation along the chain.

What would be the signature of synfire chains as seen in multi-electrode recordings of the spike activity? In such recordings the spikes emitted by few neurons picked at random are registered. If a synfire chain is triggered a certain temporal spiking pattern should be generated depending on where in the chain those neurons are located. If this synfire chain is active repeatedly, the spiking pattern should also repeat and the corresponding correlations should become visible against some background activity. Apparently such correlations have been observed with spiking patterns extending over several hundred milliseconds and with a reproducibility of less than 1 ms (Abeles, 1994). This is quite remarkable, and it requires that sufficient neurons are involved such that irregularities in the precise timing of the individual spikes are averaged out.

If the total number of neurons involved in a synfire chain or in one of its pools is small compared to the size of the total network, it is possible that several synfire chains are active at the same time. Assuming a weak coupling between different chains, synchronization of chains representing different features of the same object could be of relevance for the binding problem.

On the other hand simulations on randomly connected networks with spiking neurons and low mean activity show the existence of transients and attractors resembling synfire chains. What is typically found is a small number of long limit cycles and in addition a small number of branched long dominating transients leading into the cycles. An arbitrary initial state is quickly attracted to one of the pronounced transients or directly to one of the limit cycles. The emerging picture resembles a landscape with river systems (transients) and lake shores (cycles). It is possible that these structures serve as seeds for more pronounced synfire chains formed later by learning. It is also possible that synfire type activity is just a byproduct of other data processing events or of background activity, if such transients and attractors are always present and are not erased by learning.

5   Epilogue

The present chapter has been concerned with investigations of neural networks *as information processing devices*. The basic assumption that underlies these investigations is that information is represented by neural firing patterns, and that the spatio-temporal evolution of these patterns is a manifestation of information processing. Its course is determined by the synaptic organization of a net, which can itself evolve on larger time scales through learning. Neural networks are thus dynamical systems on (at least) two levels – that of the neurons and that of the synapses.

For higher vertebrates, there is some evidence that both speed and reliability of neural 'computations' are achieved by their being performed in *large* networks employing a high degree of parallelism. This makes up for the relatively slow dynamics of single neurons, and it gives rise to a remarkable robustness of network-based computation with respect to malfunctioning of individual neurons or synaptic connections.

The fact that we are dealing with large systems when trying to understand neural information processing suggests that concepts of statistical physics might provide useful tools in such an endeavor. This, indeed, proves to be the case, likewise on (at least) two levels – for the analysis of *neural* dynamics and associative memory, and for the analysis of the *synaptic* dynamics associated with learning and generalization.

The robustness of neural information processing with respect to various, even quite severe kinds of malfunctioning at a microscopic level – mentioned above as an observational fact – shows that microscopic details may be varied in such systems without necessarily changing their overall properties. This is to be seen as a *hint* that even quite simple models might capture the *essence* of certain information processing mechanisms without necessarily being faithful in the description of all details.

Conversely, the analysis of simplified models *reveals* that information processing in neural networks *is* robust with respect to changing details at the microscopic level, be they systematic or random. For example, the main feature of the Hopfield model (1982), viz. to provide a mechanism for associative information retrieval at moderate levels of loading, has been found to be insensitive to a wide spectrum of variations affecting virtually all characteristics of the original setup – variations concerning neural dynamics, learning rules, representation of neural states, pattern statistics, synaptic symmetry, and more. Similarly, the ability of neural networks to acquire information through learning and to generalize from examples was observed to be resilient to a large variety of modifications of the learning mechanism.

We should not fail to point out once more that the statistical approach to neural networks can claim strength and predictive power *only* in the description of macroscopic phenomena emerging as cooperative effects due to the interaction of many neurons, either in unstructured or in homogeneously structured networks. We have indicated that a number of interesting information processing capabilities belong, indeed, to this category. Our ability to analyze them quantitatively has been intimately related to finding the proper macroscopic level of description, which by itself is almost tantamount to finding the proper questions to be addressed in understanding various brain functions.

In concentrating on specific brain functions, mechanisms, and processes realizable in specific unstructured or homogeneously structured architectures, we had to leave untouched the question of how these various functions and processes are being put to work simultaneously in a real brain – supporting each other, complementing each other, and communicating with each other in the most intricate fashion. A central nervous system is after all *not* an unstructured or homogeneously structured object, but rather exhibits rich structures on many levels, with and without feedback, with and without hierarchical elements. Analyzing the full orchestration of neural processes in this richly structured system is currently far beyond our capabilities – not in small part perhaps due to the fact that we have not yet been able to discover the proper way of looking at the system as a whole.

Whether, in particular, the emergence of the 'self' will eventually be understood through and as an orchestration of *neural processes*, we cannot know. In view of the richness of the phenomena already observed at the level of simple, even primitive systems, we see no strong reason to exclude this possibility.

References

Abeles, M. (1991): *Corticonics: Neural Circuits of the Cerebral Cortex* (Cambridge University Press, Cambridge)

Abeles, M. (1994): *Firing Rates and Well-Timed Events in the Cerebral Cortex*, in: *Models of Neural Networks II*, edited by E. Domany, J.L. van Hemmen, and K. Schulten (Springer, Berlin, Heidelberg) pp. 121–141

Amit, D.J., Gutfreund H., and Sompolinsky H. (1985): *Spin-Glass Models of Neural Networks*, Phys. Rev. A **32**, 1007; *Storing Infinite Numbers of Patterns in a Spin-Glass Model of Neural Networks*, Phys. Rev. Lett. **55**, 1530

Amit, D.J. (1989): *Modeling Brain Function – The World of Attractor Neural Networks* (Cambridge University Press, Cambridge)

Amit, D.J., Brunel N. and Tsodyks, M.V. (1994): *Correlations of Cortical Hebbian Reverberations: Experiment vs. Theory*, J. Neurosci. **14**, 6445

Amit D.J. and Brunel N. (1997): *Dynamics of a Recurrent Network of Spiking Neurons Before and Following Learning*, Network, in press

Bethge, A., Kühn R. and Horner, H. (1994): *Storage Capacity of a Two-Layer Perceptron with Fixed Preprocessing in the First Layer*, J. Phys. A **27**, 1929

Braitenberg, V. and Schüz, A. (1991): *Anatomy of the Cortex* (Springer, Berlin, Heidelberg)

Cover, T.M. (1965): *Geometrical and Statistical Properties of Systems of Linear Inequalities with Applications in Pattern Recognition*, IEEE Trans. Electr. Comput. **14**, 326

Derrida, B., Gardner E. and Zippelius A. (1987): *An Exactly Soluble Asymmetric Neural Network Model*, Europhys. Lett. **4**, 167

Eckhorn, R., Bauer, R., Jordan, W., Brosh, M., Kruse, W., Munk, M. and Reitboeck, H.J. (1988): *Coherent Oscillations: A Mechanism for Feature Linking in the Visual Cortex?*, Biol. Cybern. **60**, 121

Engel, A. (1994): *Uniform Convergence Bounds for Learning From Examples*, Mod. Phys. Lett. B **8**, 1683

Gardner, E. (1987): *Maximum Storage Capacity of Neural Networks*, Europhys. Lett. **4**, 481

Gardner, E. (1988): *The Space of Interaction in Neural Network Models*, J. Phys. A **21**, 257

Gray, C.M. and Singer, W. (1989): *Stimulus–Specific Neural Oscillations of Cat Visual Cortex*, Proc. Nat. Acad. Sci. U.S.A. **86**, 1698

György, G. and Tishby, N. (1990): *Statistical Theory of Learning a Rule*, in: *Neural Networks and Spin Glasses*, edited by W.K. Theumann and R. Koeberle (World Scientific, Singapore) pp. 3–36

Hebb, D.O. (1949): *The Organization of Behavior* (Wiley, New York)

Hertz, J., Krogh, A. and Palmer, R.G. (1991): *Introduction to the Theory of Neural Computation* (Addison–Wesley, Redwood City)

Herz, A., Sulzer, B., Kühn, R. and van Hemmen ,J.L. (1989): *Hebbian Learning Reconsidered: Representation of Static and Dynamic Objects in Associative Neural Nets*, Biol. Cybern. **60**, 457

Hopfield, J.J. (1982): *Neural Networks and Physical Systems with Emergent Collective Computational Abilities*, Proc. Nat. Acad. Sci. U.S.A. **79**, 2554

Horn, D. and Usher, M. (1989): *Neural Networks with Dynamical Thresholds*, Phys. Rev. A **40**, 1036

Horn, D., Sagi, D. and Usher, M. (1991): *Segmentation, Binding and Illusory Conjunctions*, Neural Comput. **3**, 510

Horner, H. (1987): *Dynamics of Spin Glasses and Related Models of Neural Networks*, in *Computational Systems – Natural and Artificial*, edited by H. Haken (Springer, Berlin, Heidelberg) pp. 118–132

Horner, H., Bormann, D., Frick, M., Kinzelbach H. and Schmidt, A. (1989): *Transients and Basins of Attraction in Neural Network Models*, Z. Phys. B **76**, 381

Horner, H. (1992): *Dynamics of Learning for the Binary Perceptron*, Z. Phys. B **86**, 291

Jaynes, E.T. (1979): *Concentration of Distributions at Entropy Maxima*, reprinted in: *E.T. Jaynes – Papers on Probability, Statistics and Statistical Physics*, edited by R.D. Rosenkrantz (1983) (D. Reidel, Dordrecht) pp. 315–330

Kohonen, T. (1982): *Selforganization of Topologically Correct Feature Maps*, Biol. Cybern. **43**, 59

Kohonen, T. (1989): *Self Organization and Associative Memory*, 3rd ed. (Springer, Berlin, Heidelberg)

Kühn, R. and Bös, S. (1993): *Statistical Mechanics for Neural Networks with Continuous-Time Dynamics*, J. Phys. A **26**, 831

Linsker, R. (1986): *From Basic Network Principles to Neural Architectures*, Proc. Natl. Acad. Sci. USA **83**, 7508, 8390, 8779

McCulloch, W.S. and Pitts, W. (1943): *A Logical Calculus of Ideas Immanent in Nervous Activity*, Bull. Math. Biol. **5**, 115

Minsky, M. and Papert, S. (1969): *Perceptrons* (MIT Press, Cambridge, Mass.) enlarged edition (1988)

Opper, M. and Kinzel, W. (1996): *Statistical Mechanics of Generalization*, in: *Models of Neural Networks III*, edited by E. Domany, J.L. van Hemmen and K. Schulten (Springer, New York) pp. 151–209

Palm, G. (1982): *Neural Assemblies* (Springer, Berlin, Heidelberg)

Ritz, R., Gerstner, W. and van Hemmen, J.L. (1994): *Associative Binding and Segregation in a Network of Spiking Neurons*, in: *Models of Neural Networks II*, edited by E. Domany, J.L. van Hemmen, and K. Schulten (Springer, New York) pp. 175–219

Rosenblatt, F. (1962): *Principles of Neurodynamics* (Spartan, New York)

Tsodyks, M.V. and Feigel'man, M.V. (1988): *The Enhanced Storage Capacity in Neural Networks with Low Activity Level*, Europhys. Lett. **6**, 101

Virasoro, M.A. (1988): *The Effect of Synapses Destruction on Categorization in Neural Networks*, Europhys. Lett. **7**, 293

Watkin, T.L.M., Rau, A. and Biehl, M. (1993): *The Statistical Mechanics of Learning a Rule*, Rev. Mod. Phys. **65**, 499

Willshaw, D.J., Buneman, O.P. and Longuet-Higgins, H.C. (1969): *Non-Holographic Associative Memory*, Nature **222**, 960

Problem Solving with Neural Networks

Wolfram Menzel

Institut für Logik, Komplexität und Deduktionssysteme, Universität Karlsruhe, Kaiserstr. 12, D-76131 Karlsruhe, Germany. e-mail: menzel@ira.uka.de

1 A Semi-philosophical Prelude

What it means to "solve problems in a scientific way" changes in history. It had taken a long time for the paradigm of the "rigid" – the "objective", the "clare et distincte" – to become precise (and hence fixed, in the ambivalent sense of such progress): as being identical with "formalized" or "formalizable", thus *referring to a given deductive apparatus.* This appears convincing. In order to be rigid in the sense that anybody else (sufficiently trained) might understand my words and symbols as they were meant, I have to fix the language and the rules of operating on symbols beforehand, and all that "meaning" means must be contained in that initial ruling. What other way could there be to definitely exclude subjective misunderstanding and failure of any kind?

But in so many cases, it is just that fixing, just that guideline of being "rigid", which turns out to cause the major part of the problem to become intractable. Not only are we confronted with situations that are unsolvable on principle, as known from the incompleteness phenomena of various kinds: Much more important for problem solving are those boring experiences of rule-based systems running into an endless search whilst humans and animals in their actual behavior succeed quickly and efficiently.

The typical situation is that, at some stage of the computation, in order to proceed "correctly" some locally important information was needed which could not be provided beforehand in the overall coding of the problem (logically or practically). Fixing the frame of discourse in advance, i.e., being "rigid" or "formal" in the traditional sense, seems in general to unavoidably cause such weaknesses. One can certainly try to extend the machinery online, but as long as this, in turn, is done in an "algorithmic" way (i.e., a way determined by previously fixed rules), the same problem appears.

This is all well known, but there now seem to be some first small steps to nibble at the dilemma. Artificial neural networks – neural networks, for short, or ANNs – are one kind of tool that appear promising.

It is not primarily their "connectionist" or "holistic" or "subsymbolic" or "emergent" or "massively parallel" nature which makes them important. It is the fact that we can partially dispense with fixing the governing rules in advance. This is because the "programs" in ANNs are such simple things, just arrays of numbers – the *weights* –, but with the decisive property that altering the program a little does not alter the overall behavior too much. This enables "learning": From some set of examples (or other limited information) some weight configuration is achieved which controls the behavior of the system sufficiently well. That configuration may, of course, be interpreted as a "rule": but not in the sense of revealing – by translation into a given system of concepts – some "inner law" which then serves as a key to problem solution. There may be other, nonequivalent networks doing the job equally well.

Neural networks contain no magic: Learning can only be successful if an essential part of the whole design has actually been performed in a preparatory phase, e.g., by choosing the right type of network and appropriate coding of the inputs. Nevertheless, even if the amount of dispensable a priori insight is relatively small, many tasks become tractable that were not so before. Future work might help to decrease the amount of " conceptual insight" needed in advance.

Problem solving with neural networks is an approach "objective" or "rigid" enough to allow computers to help us in performing the task. That this is possible, after all, seems to bear the message that

> in order to solve a problem in a scientific way, it is far from necessary to have previously extracted the determining laws in terms of a formalism fixed in advance.

Who would ever try to solve the problem of elegantly skiing down some steep mogul field by setting up and then solving online the corresponding differential equations?

The following is a report of experiences in designing neural networks in the author's neuroinformatics group at Karlsruhe University. It covers theoretical work, exploration of possible similarities to a biological system, and applications in various fields.

2   A Fast and Robust Learning Procedure

Consider problems of the classification type, in the setting of "supervised" learning: Some finite or infinite set of *patterns* (e.g., binary or real-valued vectors) is to be partitioned into classes, and a particular problem is specified by a set of selected *training patterns*, which are given together with their corresponding class names, and the goal is to classify *all* patterns as correctly as possible. Solving problems of this type by means of acyclic ("feedforward") networks has become feasible through the learning paradigm of *error back-propagation*; see Hertz, Krogh and Palmer (1991) for historical background. For all patterns in the training set, the actual output of the network is compared to the desired "target" vector, and the difference is used to adapt the weights of the network, in a backward way from output to input, and by recursively using the corrections already made. The general idea of the weight updating is *gradient descent*, i.e., proceeding in the direction of the steepest descent of error as a function of the weights, so as to hopefully find a global minimum of that function. But there are many difficulties and shortcomings in this method if applied in its "pure" form, such as

- getting stuck in local minima
- taking oscillating "zig-zag" paths
- behaving in a counter-intuitive way: Proceeding by small steps in flat areas and by big ones near steep slopes, thus eventually missing the minimum sought.

Consequently, many modifications of the original idea and many extensions have been invented; see Riedmiller (1994) for an overview. The basic idea is to determine the steps of weight updating in a flexible and adjustable way.

RProp, or *Resilient Propagation*, extends and, at the same time, simplifies that principle (Riedmiller and Braun, 1993; Riedmiller, 1994). The weight vectors are changed in steps which

- are local, i.e., depend only on the particular component concerned
- are adaptive, i.e., depend on time
- having these two properties, are quite simple: just adding "slightly memorizing constants" to the current weights
- in particular, do not depend on the size of the gradient, but only, in a coarse way, on its direction.

If t is time, w_{ij} is the weight from neuron j to neuron i, E is the error (as a function of weights), and we let, for short, $\frac{\partial E}{\partial w_{ij}}(t)$ stand for the w_{ij}-component of the gradient of E at time t, then the weight changes are

$$\Delta w_{ij}(t) = \begin{cases} -d_{ij}(t) & \text{if } \frac{\partial E}{\partial w_{ij}}(t) > 0 \\ 0 & \text{if } \frac{\partial E}{\partial w_{ij}}(t) = 0 \\ +d_{ij}(t) & \text{if } \frac{\partial E}{\partial w_{ij}}(t) < 0. \end{cases}$$

The *update values* $d_{ij}(t)$ are determined here so as to take care of the maintenance or change, as appropriate, of the sign of the derivatives $\frac{\partial E}{\partial w_{ij}}$, thus leading to a speed up or slow down of weight changing, respectively. Using two constants η^+ and η^- ($\eta^+ > 1, 0 < \eta^- < 1$), the d_{ij} are calculated as

$$d_{ij}(t) = \begin{cases} \eta^+ \cdot d_{ij}(t-1) & \text{if } \frac{\partial E}{\partial w_{ij}}(t-1) \cdot \frac{\partial E}{\partial w_{ij}}(t) > 0 \\ d_{ij}(t-1) & \text{if } \frac{\partial E}{\partial w_{ij}}(t-1) \cdot \frac{\partial E}{\partial w_{ij}}(t) = 0 \\ \eta^- \cdot d_{ij}(t-1) & \text{if } \frac{\partial E}{\partial w_{ij}}(t-1) \cdot \frac{\partial E}{\partial w_{ij}}(t) < 0 \end{cases}$$

The dependence on the values of the parameters η^+ and η^- is rather weak, we use $\eta^+ = 1.2$, $\eta^- = 0.5$ as a standard setting.

RProp has shown better performance, faster convergence, and more robustness with respect to parameter setting than existing supervised learning procedures in very many test examples. As a consequence, RProp is widely used by other neurocomputing groups as well. Problems may arise when the training set is large, noisy, and highly redundant. Future work will have to deal with that problem. Because of RProp's particular efficiency, we have applied it in almost all neural network development in our group.

3 Obtaining Neural Networks by Evolution

For a given problem, the *topology* of the solving network (the numbers of neurons, their weights, and the connection scheme) must be chosen appropriately if training is to be successful. There are many algorithms which, for the purpose of improvement, adapt the topology and sometimes in connection with the learning process.

Another possibility is trying to "breed" the networks, i.e., to make them develop by themselves under the influence of mutation and selection. This is the approach taken by genetic and evolutionary algorithms. The system ENZO, "Evolutionärer Netzwerk Optimierer" is a realization of that idea (Braun, 1994, 1995; Braun et al., 1996).

In the general area of *evolutionary algorithms*, one roughly distinguishes between *genetic algorithms* and *evolutionary strategies*, both types possessing further modifications. In order to obtain offspring, methods of the first kind use the *cross-over* operation, which combines the respective "genetic codes" – typically bitstrings - from two given specimens, the "parents". In contrast to this, evolutionary strategies mainly *mutate* single individuals, typically working on real-valued vectors representing "phenotypes". But mutation also plays an auxiliary role in the case of genetic algorithms, and likewise, in the form of *recombination*, some more general version of cross-over enters into evolutionary strategies.

ENZO is similar to evolutionary strategies, since the cross-over operation causes certain difficulties in the case of neural networks. But being particularly tailored for the evolution of ANNs, ENZO possesses its own specifying features as an evolutionary algorithm.

ENZO works as follows: For a given problem, a *fitness* function has to be defined, which should mirror both the adequacy of the network (thus minimizing an error function) and its simplicity (corresponding to generalization ability, and to speed when the net is run). Some initial choices are made, determining, for instance, the type of networks considered, the learning procedure, the size of a population, the rate and procedure for selecting networks to be mutated, the termination condition, and some technical parameters. A first population of networks is established, e.g., by choosing in a stochastic way specimens rich in structure and with small weights, but knowledge about existing good networks can also be brought in. Then the *genetic loop* is entered; it consists of the following steps:

- From the actual population, the subset of those networks to be mutated is separated.
- Mutation is carried out. There are several mutation operators, such as removing or inserting connections, removing or inserting whole neurons together with their connections (where some local actions guarantee that the resulting changes are not too drastic).
- The offspring are trained and, after that, are incorporated in the population. Thus, some sort of "Lamarckism" is realized: Properties achieved by training can be inherited.
- Some documentation is made in order to protocol and later on evaluate the development.
- The networks are evaluated according to their fitness and, by removing the least fit ones, the original size of the population is re-established. It is advantageous to do this in a "soft" way where, with some smaller probability, less good networks can also survive.

The loop is executed until some termination criterion is fulfilled.

We have tested ENZO in problem-oriented developments of various kinds, comparing the produced networks to the best ones known from the literature for the problem concerned. ENZO's networks were much smaller, typically one third to one tenth of the size, without losing performance. There is of course a trade-off between different possible goals of evolution, such as performance and simplicity. Three of the benchmark problems will be briefly described, see Braun(1995), Braun et al. (1996).

In the "*T-C* Problem", a " *T* " or a " *C* " is to be recognized on a 4×4 pixel screen, from an input with noise and in any position on the screen. This is a small but instructive example, with combinatorial appeal. ENZO not only reduced the smallest network known from 23 to 11 neurons and from 60 to 18 weights, but also discovered that from the 16 pixel input vector only 8 pixels are actually necessary to guarantee complete success.

In order to deal with a more "real-world" example, ENZO was set the task of developing a network for recognizing hand-written numerals. A data base of 220 000 patterns was used, and the network resulting was to be compared with a commercially developed polynomial classifier, which needs 8 610 coefficients to achieve a 0.94% error (proportion of misclassifications) on the test set, the coefficients of the polynomial correspond to the weights of the network. ENZO produced several nets, e.g., one with 2 047 weights and an error of 1.14%, and one with 3 295 weights and a 0.84% error.

In a third example, a network for the classification of "half-syllables" in the context of acoustic language processing had to be developed. Because of the large amount of disturbances, this problem is much harder than the previous one. There existed a network obtained by professional hand design, possessing 15 040 weights and showing an error of 10.85%. ENZO produced a network with 893 weights and error 9.8%, additionally reducing the number of inputs required by more than half.

ENZO is being further developed to become applicable to further types of ANNs and other learning paradigms, and to deal with *feature extraction*, i.e., coding the input in an appropriate way for the given problem.

4 Models for the Olfactory System of the Honey-Bee

Honey-bees on their way to harvest nectar from blossoms navigate perfectly. Beside visual signals, odors are the main clue to the right orientation. Odors arise in permanently changing concentrations and mixtures, blown and whirled around by the wind. Contrary to visual or auditory data, they cannot be described by means of some few dimensions, such as amplitude, frequency, etc. Rather than turning one into the other in a continuous way, odors develop in plumes that break down into small odor packages with steady concentrations. Being able to manage such an abundant sensory world, the animal's odor processing system is found to be extremely adaptive, almost any odor can be learned to be "good", i.e., to indicate sugar.

In a joint project with the Institut für Neurobiologie, Freie Universität Berlin, we are investigating models for the olfactory system of the honey-bee. Honey-bees have been the favorite research object at this institute for 20 years (Hammer and Menzel, 1995; Menzel et al., 1995). It is an enormously difficult and time consuming task to supply the data needed for appropriate modeling, and, in addition to the group in Berlin, we are indebted to W. Getz, Berkeley, for supplying data. Data come, for instance, from electrophysiological recordings from individual cells, from optical recordings, and from behavioral experiments.

Odor molecules are chemically bound by the *olfactory receptor neurons* located on the antennae of the bee. Each of the two antennae carries around 30 000 such neurons. In several steps, using messenger proteins, receptor neurons transform the chemical information received into voltage spikes to be presented to the brain, hereby performing various tasks of detecting specific mixtures or changes of concentration. It has been possible to model the reaction cascade in receptor neurons by ANNs in a way that quite convincingly fits the data from experiments (Malaka et al., 1995). In that modeling, each receptor neuron has to be represented by a whole ANN. From the size of the resulting ANN, it is possible to draw conclusions regarding the number of different protein types involved in the transformation of signals.

There are two hemispheres of the brain, corresponding to the pairwise organisation of sense and motor organs, and, accordingly, the components described below occur in pairs.

From the receptor neurons, signals are passed to the two *antennal lobes* (AL). These regions in the lower part of the brain are structured, in a characteristic way, into clusters of neurites, the *glomeruli*. There are around 160 glomeruli in each AL. The result of the processing by any glomerulus is passed on by *projection neurons*. There are *interneurons* to provide the connecting, both within each glomerulus and among different glomeruli. The projection neurons, on the one hand, feed signals directly into the *lateral protocerebrum*, which controls motor activity, and, on the other hand, into the input region of the so-called *mushroom bodies* (MB). These are two symmetrically arranged neuropils of characteristic shape, extremely dense packing, and parallel arrangement of the intrinsic axons, which have attracted much attention since their discovery. Their input regions, the *calyces*, contain about one third of the neurons of the whole brain (i.e., around 170 000 each) and thus might be the main site of computation in the brain. The mushroom bodies are multimodal, i.e., they gather signals of various perceptive modes, and there is a feedback from their outputs to inputs. Also, MB extrinsic neurons project into the lateral protocerebrum. There is an identified single neuron, VUMmx1, which innervates the ALs, the MBs, and the lateral protocerebra.

Our modeling follows the idea that, after odor stimuli have been preprocessed by the receptor neurons, the ALs mainly perform tasks of cluster analysis, data compression, feature extraction, and coding of temporal dependencies: partly for immediate use in behavioral control (by the lateral protocerebrum), partly in order to prepare for associative evaluation in changing contexts. Additionally, because of the access of VUMmx1 to the AL (and as is confirmed by experiments), there seem to be *some* aspects of associative (i.e., selective) learning already performed in the AL. Modeling in this direction has led to quite satisfying results and is still going on (Malaka, 1996a,b; Malaka et al., 1996). A thesis is followed that the structuring by the glomeruli somehow reflects principal component transformation (or some more general version of this coding technique). Consequently, learning in the AL would have to cope with the more context independent parts of the whole learning task, together with some associative performances, whilst context dependent updating may be done in the calyces. There is much evidence that the VUMmx1 neuron, possessing such an impressively central position, plays the role of a "reinforcer", indicating external food rewards. Analysis and modeling of the higher parts of the brain, in particular the mushroom bodies, will be further pursued in future research.

5   Dynamic Control

The general type of problem addressed here is: to lead a given, state-based system from a given state into some desired goal state. This system – henceforth called the *process* or *plant* – might be a mobile robot, a chemical process, a particular sector of product marketing, a board game, etc.; there are a huge number of tasks of this kind. From a current state of the process, the next state can be produced by taking an *action*, to be chosen among a set of available ones, but, in general, nothing or very little is known about which state might be "nearer" to the goal than other ones. Also, the trajectories resulting may be quite long so that tree searching in the near neighborhood will not be a helpful idea.

If the dynamics of the system are known and can be described by linear differential equations (or related approximation methods), *classical control theory* can be applied to provide an explicit solution. One can use *expert systems* if sufficiently expressive rules can be found – say, in an empirical way – describing the system's behavior. But there are many situations where neither an adequate analytical modeling nor a feasible rule basis is at hand, so that working with ANNs appears appropriate.

Using these will be a quite simple task if we are equipped with (sufficiently many) data that contain the right choices of actions in given states. We can then train a network using these data and, by generalization, hope for a good action in unknown states, too. But this is a rather rare circumstance. Moreover, even if such data can be obtained, if sufficiently good and informative "experts" are at our disposal, the trained network will hardly surpass its "teachers'" ability. Anyhow, in realistic cases, such extensive information is the exception rather than the rule. We have therefore pursued two further approaches.

One of these uses a *reference model*. It may be possible to gain, from a specification of the desired behavior of the *whole* system (process together with its control unit), a description of ideal running, say in terms of linear differential equations. This reference model is used to generate " ideal" trajectories (Riedmiller, 1993). In a second step, then, facing the real process, an ANN is trained to follow those ideal trajectories as closely as possible by choosing appropriate actions, thus acting as control unit for the process.

If there seems to be no chance of finding a reference model, the only available information with which to train ANNs are rather poor *reinforcing signals*, such as "success" or " failure", after a long sequence of states. The task, then, consists in the gradual formation of an *evaluation function* of the states, by exploration of the whole state set: if such an evaluation were known, actions could always be taken such that the next state had an optimal value. This establishes what is known as the *temporal credit assignment problem*. The central question is how to build up that globally good evaluation by adjustments that use local information only. In a general, theoretical framework, the problem has been solved by Dynamic Programming (Bellman, 1957). But it is only in special cases that the method provides an algorithm finding the evaluation function, e.g. if the state transition function is known and is linear. We have tried to cope with the general situation by using *reinforcement learning* in the form of the temporal difference (TD) approach, (Sutton, 1988; Watkins, 1989). Here, estimations for the values of the states are gradually improved by propagating final ("reinforcing") signals back through trajectories, in each step using only local information (Riedmiller, 1996; Janusz and Riedmiller, 1996).

The approaches mentioned have been tested for many examples, some in simulations and some in reality, in most cases quite successfully. These examples include

- catching a rolling ball
- controlling a little mobile robot so as to avoid obstacles
- balancing a cart pole (in a real world, simple, and " dirty" setting)
- balancing a double cart pole (simulation)
- gas control in combustion engines
- some board games.

Most of these are problems in a continuous world (Riedmiller 1996, 1997; Janusz and Riedmiller, 1996) but the problems of the last type use discrete state spaces (Braun et al., 1995).

There seems to be a wide variety of tasks of dynamic systems control that can be appropriately solved by ANNs, but appear inadequate for analytical or rule-based approaches. Difficulties still arise in cases of huge discrete state spaces, as associated, e.g., with chess. More powerful methods will have to be developed for such types of problems.

6   Finding Structure in Music

Finding structure in music, i.e., finding guidelines that govern good composing, has been the subject of music theory for centuries. Laws for cadences and counterpoint have been established, laws and recommendations for harmonizing a given melody, schemes of repetition and variation, and many more convincing concepts for the analysis of musical works and, finally, a partial support of composing.

But it is neither likely, nor desirable, that one could describe by abstract laws a chosen (sufficiently good) piece of music down to its individual peculiarity. Would not the very existence and knowledge of an algorithm producing convincing compositions immediately lead to a redefinition of "good", such that now breaches of the rules laid down beforehand appeared to be qualifying features?

Finding structure in music seems to have its own specific dialectics, its limitations, gaps, and perhaps paradoxes. The question arises of whether neural networks can perform more than rule-based systems can, or can cover different aspects, from the task of "objectivizing" musical events, namely of making accessible such events to non-human processing. We have begun to experiment on this question.

Our best developed example is HARMONET, a network which finds a four part harmonization to a given melody, in the choral style of J. S. Bach (Hild et al., 1992). (It requires a one to two year education for an organist to acquire this ability.) For this task, rule-based systems have been developed, e.g. the rich expert system CHORAL by Ebcioğlu (1986). As opposed to such approaches, which are necessarily specific for some selected musical style, HARMONET gains its ability from training by examples and can thus be retrained for other composers and styles.

HARMONET works roughly as follows. The training examples, which are original harmonizations by Bach, have to be prepared in a special way. Ornamentations are omitted, and the remaining chords are replaced by their class names (such as "tonica", "tonica-parallel", "dominant-septime") yielding the *harmonic skeleton*. The choral is then cut into pieces, in order to

Christus, der ist mein Leben

Nicht so traurig, nicht so sehr

Fig. 1. Two examples of harmonizations found by HARMONET

allow for local training: Given a short past of harmonies already found and the next two melody tones, the net is trained to predict the chord class name belonging to the next note, by comparison of its choice to that by Bach. After the harmonic skeleton has been settled, the chords themselves are formed. This can be done by either a rule-based system or a neural network. Finally ornaments are inserted; a particular ANN has been trained for this subtask.

A crucial point for HARMONET's success was the coding of the pitches. Our choice was to code them as *harmonic functions*, i.e., a pitch mirrors the chords it belongs to.

Figure 1 shows two examples of harmonizations found by HARMONET, one in a major key and one minor. These are, of course, proper examples, i.e., the chorals did not belong to the training set. The letters below the lines indicate the harmonic skeleton.

We are currently investigating the formation of melodies. This has turned out to be a problem very much harder than harmonizing, but there are encouraging first results. An ANN has been developed that realizes the playing around a given melody, in sixteenth notes, in the style of J. Pachelbel (Feulner and Hörnel, 1994). The network consists of two parts connected with each other, one for the coarser planning, the finding of "motifs", and one for implementing those motifs by particular notes. The results produced by the network are impressive (Hörnel and Degenhardt, 1997). Future work will be on topics such as melody development in folk songs, or finding a supporting voice.

7   Prediction of Financial Movements

It is tempting to try to use neural networks for predicting the development of economic data, such as foreign exchange rates or stock prices; the fantasy of marvellous profits is always waiting round the corner. These, indeed, are very unlikely to occur, as some few reflexions on the topic will reveal. But we have nevertheless been able to develop a network for foreign exchange trading which shows remarkable performance. The project reported in the following is a collaboration with Helaba, the State Bank of Hessen and Thüringen in Frankfurt.

The data underlying our analysis have been the daily exchange rates (high/low/close) US\$/DM from the New York stock exchange, first from January 1986 to June 1992, later on up to the present. Corresponding data from the Frankfurt stock exchange have also been made available. We have additionally been supplied with information about some technical indicators of the type frequently used by traders, such as moving averages and other clues to a specific trend.

It is natural to consider the development of exchange rates as a stochastic process. Modeling the (hoped for) statistics behind some given time series of data will become rather straightforward if the correlation found in the data is significant, thus showing a linear dependency between successive values. But correlation was negligible in our data. It could then have been the case that the process was just a random walk, with no dependency at all of the current exchange rate on former ones. As analysis showed, this is also not the case. The impression resulting then is that there are weak non-linear dependencies between the data, with strong superposed noise. This is in accordance with results from the literature. A further statistical analysis, then, tends to become quite complicated, and it appears natural to apply neural networks.

One might try to train an ANN just by the samples given, using (e.g., closing) exchange rates from n successive days as input and that one from the following day – or a coarsened version of it, such as an interval, or a ternary value rise/stay/drop – as a target value to be approximated by the output. It quickly becomes apparent that such a simple approach, or similar ones, will not work: The results are hardly better than the outcomes of just throwing dice would be. Further analysis is necessary.

The main difficulty in our problem is that of the "single line". Suppose we had at our disposal some thousands of realizations of the (identical) statistical rules governing the US$/DM market, each one as a time series, differing from each other by slight modifications ("disturbances") in accordance with those rules. Then, possessing a rather rich ensemble of sample series, we would be able to apply estimation techniques, or could select particular days where disturbances appear to be minor: Using these days for training, we could be more confident that a good ANN predictor was attainable.

But we do not have those thousands of parallel market realizations, there is just one. All we can do is trying to overcome that lack of data by tricks, such as smoothing the curve in various ways; coarsening the output; inventing and adding "indicators" (which are quantities hopefully not so afflicted by noise as the data themselves); inventing criteria to select "expressive" days (where the exchange rate should be more strongly "rule-governed" than on others). As, in general, we are unable to prove that any step taken in that direction will work, finding successful arrangements is, in turn, an empirical task, and has been integrated in various ways in the development of ANNs. Thus, even if a "good" design seems to have been found, the trained network might still mirror peculiarities of the market which were present in the training period, but need not be in future times. A property very desirable would be the ability to distinguish between certain "types" of market, but indicators for those types can hardly be found. A good method to overcome difficulties of this latter kind is using a competitive team of ANNs ("experts"), each of them having formed its own model of the market.

It took around two years of analytical and empirical work to develop a rather successful neural network for the US$/DM exchange rate prediction (Gutjahr et al., 1997). The output values are binary recommendations, "rise" or "drop", for daily trading. Averaging over six years, around 54.5% of the

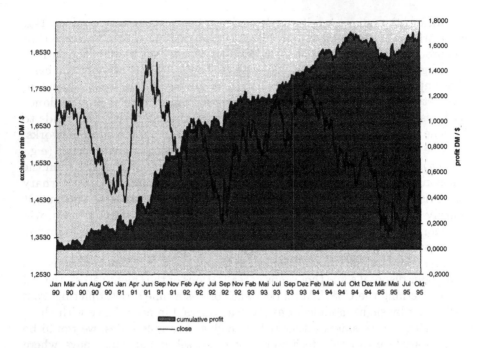

Fig. 2. Cumulative profit made by predicting, on a daily basis, the US\$/DM exchange rate

predictions have been correct. From January 1990 to April 1996, the total gain has been DM 1.673 per dollar invested in the trading (Fig. 2).

Since July 1995, the network has been used by the Helaba for factual trading at New York stock exchange. Appreciation by its human colleagues has been expressed by baptizing it "George".

We are trying to further improve George. One essential guideline will be to use ENZO as a more systematic method of developing teams of good predictors.

There are many more sections of the financial market that may successfully be tackled with neural networks. Thus, an ANN for trading with futures has recently been developed and is now being applied by the Helaba (Gutjahr, 1997).

8   Conclusions

Artificial neural networks have turned out to be powerful tools for solving problems, in particular in real-life situations. Difficulties and deficits in the preparation of a question can be partially overcome by the adaptivity of the ANN mechanism.

In our activities reported here, analytical and empirical work had to go hand in hand, and was simultaneously challenged by the modeling of a complex natural system. As has been demonstrated, the range of problems that can successfully be tackled with neural networks is enormous. Another experience is that empirical investigation appears as the natural and compelling continuation of formalizing and analytical approaches, rather than something antagonistic to them, or some subordinate companion.

Our research originated from questions of logic and tractability. It has been the declared goal of investigating neural networks to take as constructive the well-known, seemingly "negative" or "hindering" phenomena of incompleteness and intractability. As a matter of fact, these *are* constructive phenomena.

Acknowledgements

It is gratefully acknowledged that the reported work was, in parts, supported by

> Deutsche Forschungsgemeinschaft
> Land Baden-Württemberg
> Landesbank Hessen-Thüringen
> Daimler Benz AG.

Beside the author, our group consists of: H. Braun, St. Gutjahr, D. Hörnel, R. Malaka, Th. Ragg, M. Riedmiller. The author wishes to thank them for reading this article and commenting upon it.

References

Bellman, R. E. (1957): *Dynamic Programming* (Princeton University Press, Princeton, New York)

Braun, H. (1994): *Evolution – a Paradigm for Constructing Intelligent Agents*, Proceedings of the ZiF-FG Conference Prerational Intelligence – Phenomenology of Complexity Emerging in Systems of Simple Interacting Agents

Braun, H. (1995): *On optimizing large neural networks (multilayer perceptrons) by learning and evolution.* International Congress on Industrial and Applied Mathematics ICIAM 1995. Also to be published in Zeitschrift für angewandte Mathematik und Mechanik ZAMM

Braun, H., Feulner, J., and Ragg, Th. (1995): *Improving temporal difference learning for deterministic sequential decision problems* International Conference on Artificial Neural Networks (ICANN 1995), (Paris, E2 and Cie) pp. 117–122

Braun, H., Landsberg, H. and Ragg, Th. (1996): *A comparative study of neural network optimization techniques,* Submitted to ICANN

Ebcioğlu (1986): *An expert system for harmonization of chorales in the style of J. S. Bach,* Ph.D. thesis, State University of New York (Buffalo N.Y.)

Feulner, J. and Hörnel, D. (1994): *MELONET: Neural networks that learn harmony-based melodic variations*, in: Proceedings of the International Computer Music Conference (ICMC) (International Computer Music Association, Århus)

Gutjahr, St. (1997): *Improving neural prediction systems by building independent committees*, Proceedings of the fourth International Conference on Neural Networks in the Capital Market (Pasadena, USA)

Gutjahr, St., Riedmiller, M. and Klingemann, J. (1997): *Daily prediction of the foreign exchange rate between the US dollar and the German mark using neural networks*, The Joint 1997 Pacific Asian Conference on Expert Systems/Singapore International Conference on Intelligent Systems (1997)

Hörnel, D. and Degenhardt, P. (1997): *A neural organist improvising baroque-style melodic variations*, Proceedings of the International Computer Music Conference (ICMC) (International Computer Music Association, Thessaloniki)

Hammer, M. and Menzel, R. (1995): *Learning and memory in the honeybee*, J. Neuroscience 15, pp. 1617–1630

Hertz, J., Krogh, A., and Palmer, R. (1991): *Introduction to the Theory of Neural Computation* (Addison-Wesley, New York)

Hild, H., Feulner J. and Menzel, W. (1992): *HARMONET: A neural net for harmonizing chorals in the style of J. S. Bach*, in: Advances in Neural Information Processing 4 (NIPS4), pp. 267–274

Janusz, B. and Riedmiller, M. (1996): *Self-learning neural control of a mobile robot*, to appear in: Proceedings of the IEEE ICNN 1995 (Perth, Australia)

Malaka, R. (1996): *Neural Information processing in insect olfactory systems*, Doctoral dissertation (Karlsruhe)

Malaka, R. (1996): *Do the antennal lobes of insects compute principal components?*, Submitted to: World Congress on Neural Networks '96 (San Diego)

Malaka, R., Ragg, Th., and Hammer, M. (1995): *Kinetic models of odor transduction implemented as artificial neural networks – simulations of complex response properties of honeybee olfactory neurons*, Biol. Cybern., 73, pp. 195–207

Malaka, R., Schmitz, St., and Getz, W. (1996): *A self-organizing model of the antennal lobes*, Submitted to the 4th International Conference on Simulation of Adaptive Behaviour, SAB

Menzel, R., Hammer, M., and U. Miller (1995): *Die Biene als Modellorganismus für Lern- und Gedächtnisstudien*, Neuroforum 4, pp. 4–11

Riedmiller, M. (1993): *Controlling an inverted pendulum by neural plant identification*, Proceedings of the IEEE International Conference on Systems, Man and Cybernetics (Le Touquet, France)

Riedmiller, M. (1994): *Advanced supervised learning in multi-layer perceptrons – From backpropagation to adaptive learning algorithms*, Computer Standards and Interfaces 16, pp. 265–278

Riedmiller, M. (1996): *Learning to control dynamic systems*, to appear in: Proceedings of the European Meeting on Cybernetics and Systems Research (EMCSR 1996) (Vienna)

Riedmiller, M. (1997): *Selbständig lernende neuronale Steuerungen*, Dissertation Karlsruhe 1996, Fortschrittberichte VDI-Verlag (Düsseldorf)

Riedmiller, M. and Braun, H. (1993): *A direct adaptive method for faster backpropagation learning: The RPROP algorithm*, Proceedings of the IEEE International Conference on Neural Networks (ICNN) (San Francisco), pp. 586–591

Sutton, R. S. (1988): *Learning to predict by the methods of temporal differences*, Machine Learning, 3, pp. 9–44

Watkins, C. J. (1989): *Learning from delayed rewards*, Ph.D. thesis, Cambridge University.

Paradigm Shifts
in the Neurobiology of Perception

Andreas K. Engel[1] and Peter König[2]

[1] Max-Planck-Institut für Hirnforschung, Deutschordenstr. 46,
D-60528 Frankfurt, Germany. e-mail: engel@mpih-frankfurt.mpg.de
[2] The Neurosciences Institute, 10640, John Jay Hopkins Drive,
San Diego, CA 92121, USA. e-mail: peterk@nsi.edu

Introduction

In cognitive science, we are currently witness to a fundamental paradigm shift [for review see, e.g., Varela et al. (1991) and Bechtel and Abrahamsen (1991)]. In the early days of artificial intelligence and cognitive psychology, cognitive processes were assumed to be based on algorithmic computations controlled by formalizable rules which act upon quasi-propositional knowledge about the external world. In recent years, however, more and more cognitive scientists have converted to the idea that functions like perception, problemsolving, or memory emerge from complex interactions in highly distributed neuronal networks which, unlike conventional information-processing systems, are shaped by learning and experience-dependent plasticity. In such networks, information processing does not follow explicit rules, but is based on the self-organization of patterns of activity. This paradigm shift is motivated by the apparent inability of classical models to account for many facets of cognitive processes and, in addition, by their lack of biological plausibility. In the present contribution, we describe this paradigm shift as it occurs in the neurobiology of perception, with particular reference to the visual system. In addition to demonstrating crucial differences, we elaborate basic conceptual assumptions which are common to both the classical neurobiological view on perception and to the more recent connectionist framework. We then attempt at developing a critical perspective on these key assumptions and suggest that they possibly need to be modified in several respects to account more appropriately for the phenomenology of perceptual processes.

1   The Classical Paradigm: Brains as Serial Computers

Among those pieces of work which have been most important for the foundation of modern visual neurobiology are the pioneering studies of David Hubel and Torsten Wiesel. In several respects, their investigation of the response properties of neurons in the visual cortex had a long lasting and deep impact on the view that neurobiologists have of perception (for a review of these findings, see Hubel 1988 pp. 42–43 and pp. 67–88). First, they contributed to the concept of the *receptive field* by demonstrating that neurons in the visual system respond to stimulation only in a circumscribed region of the visual field. Second, they discovered that receptive fields can be structured in certain ways and prefer certain stimuli over others (Fig. 1). This finding led to the notion that sensory neurons act as feature detectors and established the idea that patterns of neuronal activity can be considered as *representations* of objects or events in the external world. Third, they proposed that features of objects are processed in a *hierarchical* fashion, and that neurons can acquire more complex response properties by integrating the computational outputs of cells responding to simple features, such as oriented edges or moving corners.

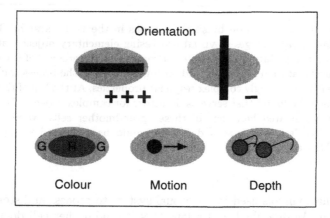

Fig. 1. Response properties of cortical neurons. A paradigmatic example is the orientation-selectivity of many cortical neurons (top). In the case illustrated here, the neuron would discharge at a high rate when a horizontally oriented stimulus (black bar) is presented in the receptive field (grey area), i.e., the portion of the visual field where the activity of the cell is influenced by light. If the orientation of the stimulus is changed, the discharge rate decreases and, eventually, complete inhibition of the neuronal activity occurs. Other typical response properties observed in cortical neurons are color selectivity (bottom, left; R, red-sensitive area; G, green-sensitive area), specificity for the direction of stimulus motion (bottom, middle) or selectivity for the relative spatial distance of the object (bottom, right).

In a seminal paper, Horace Barlow generalized this set of hypotheses to a fullfledged theory of object recognition (Barlow, 1972). He suggested that progressive convergence of the outputs of feature-detecting neurons through a hierarchy of processing stages could finally yield cells with highly specific response properties (Fig. 2). According to this hypothesis, even complex objects, such as one's own grandmother, could be represented by the activity of *single neurons*. Such "grandmother cells" would then constitute the rep-

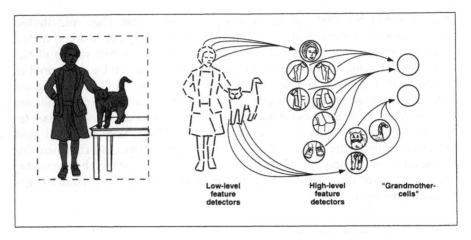

Fig. 2. Object representation by single neurons in the visual system. The model assumes that at early stages of visual processing elementary object features such as, e.g., contour orientation are detected. Progressive convergence of the outputs of these low-level feature detectors should, at higher levels in the processing hierarchy, yield cells with increasingly complex response properties. At the top of the hierarchy neurons should be found that serve as detectors for complex objects. In case of the visual scene illustrated here, one of these "grandmother cells" would signal the presence of the lady and a second neuron would be dedicated to represent the presence of the cat.

resentational symbols used by the visual system to process and store sensory information. As described by Barlow, this "grandmother cell doctrine" has several important conceptual implications: First, information processing is assumed to be strictly *serial* and to occur in a purely feed-forward manner. In addition, processing is highly *localized* and, at least in higher processing stages in the visual system, only a few specialized cells are assumed to be active at any time. Second, individual neurons are considered as relatively complex computational units. Being the substrate of mental representations, they constitute the "semantic atoms" of brain activity, whose meaning is *independent* of the activity of other neurons at the same processing level. Third, according to this framework cognitive processes are tightly correlated with the behavior of single neurons. Barlow postulated that activation of the respective "grandmother cells" should be both necessary *and sufficient* to account for the occurrence of phenomenal states and the perceptual *experience* of a certain object. For these reasons, he concluded, studying the firing patterns of single neurons was doubtlessly the relevant level of description for both neurobiology and psychology. By formulating this "grandmother cell doctrine", Barlow made explicit some of the key assumptions which deeply pervaded the thinking of neurobiologists at this time. The picture that Barlow and his colleagues drew of brain processes and of the neural correlates of perception seemed to tie up neatly with the metaphors developed by artificial intelligence researchers: the brain was considered as a serial computer that embodies logical principles in its architecture and calculates symbolic representations according to explicit algorithms.[1]

2 The Connectionist Paradigm: Self-Organization in Neuronal Networks

Although the classical paradigm just described still shows considerable persistence in the minds of neurobiologists, this view of the brain has been challenged in recent years. Increasing evidence suggests that the visual system performs its task in an entirely different way and that information processing is not carried out as a sequence of discrete computational steps. In addition, neurobiologists failed to disclose the "grandmother neurons" advocated by Barlow, and it turned out that presumptive "representational" brain states look entirely different. Finally, it became undeniably that the computer metaphor is neither a helpful nor a valid analogy, because brains do not show anything like a central processing unit, they do not contain any obvious "rules" or algorithms, and – unlike in von-Neumann machines – the same parts of the system which "compute" also provide the substrate for memory. These insights have inspired a new, connectionist picture of perceptual processing, which differs from the classical framework with respect to several crucial assumptions and highlights a number of new concepts.

Parallel and distributed processing: By the mid-eighties it had become clear that Hubel's and Wiesel's classical notion of processing in a hierarchical chain of visual areas was no longer tenable. Instead, evidence was accumulating that neurons respond to visual stimuli in a large number of areas, which seem to be specialized for processing different attributes of visual objects, such as their form, color, location in depth, or motion trajectory.[2] Rather than being part of simple hierarchy, these areas form a complex *network* with numerous input and output pathways, in which each node is linked to its neighbors by reciprocal connections (Fig. 3). In this network, information is processed at many sites *simultaneously* and – contrary to Barlow's intuition –

---

[1] These ideas were developed, among others, by McCulloch and Pitts, and by Newell and Simon in their work on the "physical symbol system hypothesis". See Varela et al. (1991) pp. 39–48 for further references and for a review of this early phase of cognitive science. This view is also implied in Marr's classical approach to vision and object recognition (Marr, 1982).

[2] To date, more than 30 areas have been identified in the monkey visual system. These observations, together with the discovery that neurons with quite different response properties are already present in the retina and subcortical visual structures has led to the notion of "parallel processing streams". For review, see Felleman and Van Essen (1991), Spillmann and Werner (1990) pp. 103–128 and pp. 194–199.

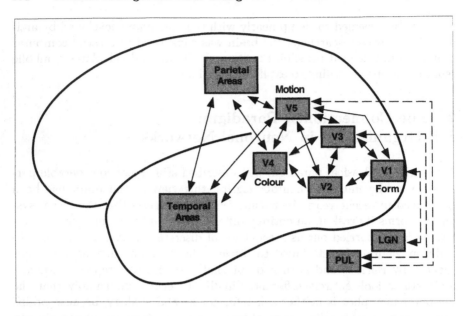

Fig. 3. Parallel and distributed processing in the visual system. Only a fraction of the visual areas known to date are depicted in this schematic diagram (grey boxes). As indicated by the arrows, almost all connections between these areas are reciprocal. Is is generally assumed that different areas are, at least to some degree, specialized for the processing of different object features such as, e.g., form, color or motion. Note that the cortical system has multiple parallel links to subcortical structures. V1 ... V5, first ... fifth visual area; LGN, lateral geniculate nucleus; PUL, pulvinar.

large numbers of neurons distributed throughout the network are coactivated by each object present in a visual scene.

Self-organization and plasticity: Both the ontogenetic development of this huge cortical network and the changes of its topology during adulthood are now considered as processes of self-organization during which *local* rules give rise to *globally* ordered patterns (for a review, see e.g. von der Malsburg and Singer, 1988). Thus, for instance, the pattern of axonal connections within and between cortical areas is not genetically pre-programmed but is largely determined by the action of local correlation rules. In addition, there is now ample evidence that such connectivity patterns are shaped by the behavior and the actual sensory experience of the animal. Even in the adult state, these networks show a high degree of activity-dependent plasticity, which is related to functions like learning and memory (see Singer, 1995).

Context-dependence of neuronal responses: Contrary to what Barlow assumed, it now seems clear that the significance of a neurons' firing cannot be considered in isolation. In at least two ways, neuronal responses are highly

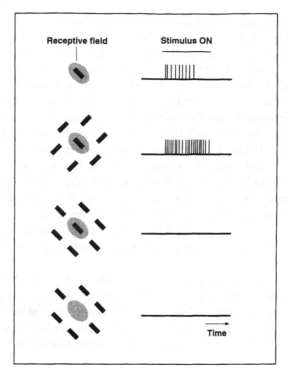

Fig. 4. Context-sensitivity of cortical neuronal responses as observed in the primary visual area. The response to a stimulus projected into the receptive field proper (grey area) is strongly modulated by other stimulus elements in the vicinity of the receptive field area. The neuron increases its firing rate upon presentation of an optimally oriented stimulus (top trace). If additional stimulus elements of orthogonal orientation are shown outside the receptive field, the response is strongly enhanced (second trace), but inhibition can occur if these line elements have the same orientation as the stimulus placed into the receptive field (third trace). This modulation occurs although the surround stimuli as such do not elicit a response from the neuron (bottom trace).

dependent on the context in which they occur. First, it has been shown that the firing rate of a given sensory neuron is not only influenced by the stimulus present in its receptive field. Rather, the response depends on the *whole* of the stimulus configuration and on the way in which neighboring regions in the visual field are affected (Fig. 4).[3] In addition, neuronal responses are strongly modulated by the behavioral context and the attention of the experimental animal [for a review of these modulatory effects see, e.g., Spillmann and Werner (1990) pp. 220–225 and pp. 308–313]. Second, there is now evidence that responses of individual neurons are not per se causally efficacious

---

[3] This effect has been described, e.g., by Knierim and Van Essen (1992).

elsewhere in the brain. Single neurons influence behavior and, thus, acquire "meaning" only if they are part of coherently active assemblies.

Assemblies as basic functional units: Recent evidence indicates that *coherently active populations* of cells are the meaningful functional entities of information processing (Fig. 5), rather than single neurons as postulated by Barlows "grandmother cell doctrine". In contrast to grandmother neurons, for which experimental proof is still lacking, linkage of cells into assemblies has now been directly observed; for review, see Engel et al. (1992) and Singer and Gray (1995). Using simultaneous recordings with multiple electrodes, it has been shown that spatially separate cells in the visual cortex can synchronize their firing if they respond to the same object present in the visual field. By response synchronization, assemblies can be formed even across different visual areas and across the two cerebral hemispheres. This "binding" of distributed neurons into assemblies, which is mediated by local interactions, provides another example for the self-organization of ordered patterns in neuronal networks.

Dynamics of neural processing: Finally, the connectionist viewpoint emphasizes the notion that cognitive systems are dynamic, i.e., their operating principles can only be captured if one considers their evolution in time. Technically, this implies the usage of dynamical systems theory and the introduction of concepts such as phase spaces and attractors into the neurobiological

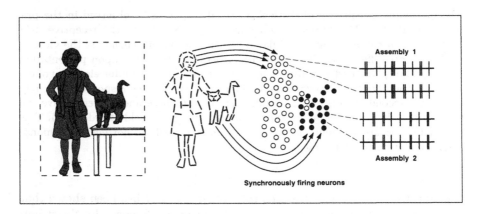

Fig. 5. Object representation by distributed cell assemblies. This model assumes that objects are represented in the visual cortex by assemblies of synchronously firing neurons. Thus, the lady and her cat would each be represented by one such assembly (indicated by open and filled symbols, respectively). These assemblies comprise neurons which detect specific features of visual objects (such as, for instance, the orientation of contour segments) within their receptive fields. The relationship between the features is encoded by the temporal correlation among the discharges of these neurons (right). The model assumes that neurons which are part of the same assembly fire in synchrony whereas no consistent temporal relation is found between cells belonging to different object representations.

description.[4] Examples of how the notion of dynamics influences perceptual neurobiology are provided by current work on the structure of receptive fields and on the nature of representational states in the visual system. Thus, recent evidence shows that the spatial structure of receptive fields in central visual neurons evolves in time after the stimulus has appeared and thus, these fields have to be characterized as spatiotemporal entities without the possibility of separating spatial and temporal components; see, e.g., DeAngelis et al. (1995). Very similar considerations hold for the structure of representational states which, as discussed above, can be established by the transient and flexible synchronization of feature-detecting neurons. In contrast to this connectionist view, the classical framework had largely neglected the dynamical aspects of neural systems.

Taken together, these concepts clearly challenge the view of the brain developed in the classical framework, and argue for a much more dynamic and holistic picture. In the new framework, information processing is not described as algorithmic, but as a spread of activity through cortical areas leading to self-organization of highly distributed spatiotemporal patterns which represent the computational "result". These patterns are determined by the topology of the cortical network and by its history during which learning has induced adaptive structural changes. Interestingly, these conclusions imply a return to central insights of the Gestalt psychologists. Although the various brain areas are by no means "equipotential", processing is never strictly localized and cognitive processes are always accompanied by activity in large parts of the brain. And indeed, brain states seem to have a gestalt-like organization, since the functional significance of the activity of individual neurons depends on the context set by other members of the neural assembly.

3 Towards a New Framework for the Neurobiology of Perception

Although the two paradigms described above differ considerably in their view of how information is processed and stored in the nervous system, they share a set of fundamental assumptions. First, the defenders of both paradigms adhere to a *representational theory of perception*, i.e., they assume that perception consists in recovering features of a pre-given world and in constructing a copy, or internal image, of this world in the perceptual apparatus. Second, these paradigms share an *atomistic ontological framework* as well as *atomistic methodological premises*. And third, in both paradigms most researchers seem to maintain the rarely questioned assumption that looking into brains is not only necessary but, in the long run, *sufficient* to arrive at an appropriate understanding of cognitive processes. We will now discuss each of these philosophical assumptions in turn.

---

[4] A recent collection of articles illustrating this point can be found in Port and van Gelder (1995).

Representational theory of perception: Both the classical and the connectionist view subscribe to ontological realism, i.e., they consider the world as existing independent of and prior to any cognitive activity. Confronted with this world of pre-defined structures, cognitive systems are essentially passive and behave in a merely receptive way. They are thought to "reconstruct" objects or events of the external world by virtue of computing representations, the latter being considered as *internalized images* on the basis of which the cognitive system produces some "motor output". Thus, to put it in a nutshell: "brains are world-modelers" (Churchland and Sejnowksi, 1992, p.143). Criticism of this view of cognition has a long tradition in modern philosophy, and a review of this ongoing debate is certainly beyond our present scope. However, we wish to suggest an alternative view which is based on a phenomenological position.[5] From a phenomenological point of view, it has been argued that such a representational framework cannot sufficiently account for perceptual and cognitive processes, because it neglects their creative, or *constructive*, aspects.[6] Being perceivers, we are literally creating a phenomenal world, because the process of perception first of all defines relevant distinctions in the sensory environment.[7] In visual perception, phenomenal items such as edges, textures or colors are always specified *relative* to the observer for whom these distinctions are relevant. Thus, the structures of the perceptual world are inseparable from the cognitive agent and, in this sense, "world-making" rather than "world-mirroring" seems to lie at the heart of cognition; see, e.g., Merleau-Ponty (1976), pp. 11–16, 138–143 and 176–183. Drawing on the phenomenological tradition, Varela has recently emphasized the role of action, which he considers as intrinsically linked with perception: Cognition, as he puts it, can be understood as the capacity of *"enacting"* a world.[8] This concept is not meant to imply that mental or neuronal representations do not exist. Rather, what this view emphasizes is that, *if* the notion of "representation" refers to creating passive mirror-images of the external world, *then* the process of representing cannot be at the core of our

---

[5] When using the term "phenomenology", we refer to the positions developed in the the early writings of Heidegger and those of Merleau-Ponty. See, e.g., Heidegger (1989) and Merleau-Ponty (1976).

[6] It should be noted that in using the notion of "constructive aspects" of cognition we are, of course, not referring to volitional acts carried out by a rational conscious ego (a "cartesian subject"). What we want to highlight is the idea that the contents of perception (and, hence, the structure of the phenomenal world) is largely determined by the self-organizing dynamics of the cognitive system and by prerational sensorimotor dispositions that are embodied in the cognitive agent. In this sense, what we argue for is "construction without a constructor".

[7] We are referring to the process of scene segmentation, i.e., the breakdown of a visual scene into meaningful chunks of information. Segmentation is based on criteria that are embodied in the architecture of the cognitive system.

[8] See Varela et al. (1991), pp. 133–145, 147–150 and 172–178 for critique of the concept of representation and the "enactive" view of cognitive processes.

cognitive capabilities. Accepting this "enactive" view necessarily leads to a redefinition of the neurobiologist's explanandum: What neuroscience, then, has to explain is not how brains act as world-mirroring devices, but how they can serve as "vehicles of world-making". Accordingly, the contents of brain states would not be information about pre-given objects or events in the world. Rather, neural states could be viewed as primarily representing the capacities of structuring situations and of creating perceptual constructs that the respective organism is endowed with.

Atomistic ontology:[9] Clearly, the classical paradigm of perception views the world as a universe of independent and context-free "features" or "objects", which are neutral with respect to the cognitive agent who enters the scene.[10] Although connectionist models constitute some advance with respect to incorporating context, they essentially share this ontology.[11] In such models, "meaning" and relevance are extracted by simply counting the incidence of "features" and by monitoring which of these frequently occur in spatial and temporal contiguity, the latter then leading to associative links in the network. Thus, perceptual learning in connectionist models merely amounts to becoming familiar with the statistics of input patterns. Therefore, despite showing holistic traits connectionism still has an impoverished understanding of "context", which is reduced to the coincidence of contingent features, and according to which relationships in the cognitive system are established in a bottom-up manner by simple correlation rules. We suggest replacing this atomistic ontology of "neutral features" by a *holistic ontological framework*.[12] According to Heidegger and Merleau-Ponty, what we encounter as cognitive agents are never "bare" objects or arrays of contingent features but, rather, meaningful situations, i.e., contexts which we have already structured by prior activity and in which objects are defined *as a function of* our needs

---

[9] We use the term "ontology" in the sense of Dreyfus (1992), i.e. referring to basic assumptions on the nature of the relevant entities in the domain of a scientific theory.

[10] Current theories on perceptual segmentation frequently assume that natural scenes have objectively defined boundaries or "predefined breaks", i.e. there is only one natural or "correct" way of segmenting a scene. Consequently, the task of the sensory system is to find the correct solution to the segmentation problem. This is implied, for instance, in the theories advocated by Marr (1982) and Biederman (1987).

[11] Therefore, the critique of Dreyfus (1992), pp. 206–224, originally aiming at the classical cognitivist paradigm, is still valid and readily applicable to connectionism.

[12] Using the notion of "holism" we refer to the view that the phenomena relevant to a theory of perception cannot by understood or predicted by studying lower-level elements in isolation, because the significance of the latter depends crucially on the context they are embedded in.

and concerns.[13] Even for the newborn, the world is not a heap of coincident features, since its own needs in concert with the social context define what the world should look like. This situational ontology implies that objects are not context-invariant entities but, rather, are individuated according to the situation's demands and according to the task at hand. Accordingly, "object representations" cannot be considered as containing invariant structural descriptions. Rather, they refer to objects as-being-embedded-in-a-situational-context.

Atomistic methodology: In addition to their atomistic ontology, both paradigms share atomistic methodological premises. As described earlier in this chapter, the proponents of the classical framework have emphasized the single neuron as the relevant level of description. In his "grandmother cell doctrine", Barlow expressed the belief that "it no longer seems completely unrealistic to attempt to understand perception at the atomic single-unit level" (Barlow, 1972, p. 382). Although this doctrine now no longer represents a majority view, the atomistic intuition still appears in many facets. Thus, for instance, it inspires the concept of *modularity* of cortical processing: "The assumption is that the visual system consists of a number of modules that can be studied more or less independently. ... The integration of modules is assumed to be primarily 'late' in nature" (Ullman, 1991, p. 310). This concept also implies that the process of vision as a whole can be treated in isolation from other cognitive processes. Thus, it is thought that the visual system can successfully solve relevant problems, such as scene segmentation and object recognition, by *exclusively* operating upon visual cues.[14] Consequently, it is assumed that objects can be represented as purely sensory patterns in the brain. Contrary to this now-popular notion, we suggest that a theory of perception remains incomplete if referring to sensory processes only. It needs to be taken into account, for instance, that perception is always part of ongoing activity of the organism. Thus, seeing an object does not correspond to purpose-free extraction of its visible features but, rather, to visually guided action in a certain situational context.[15] This view implies that our knowledge of objects does not rely on abstract structural descriptions derived from sensory

---

[13] Introducing the notion of "situation", we are aiming at what Heidegger has termed the "Bewandtnisganzheit", see e.g. Heidegger (1989), p. 231–242. Along similar lines, Dreyfus points out that "a normal person experiences the objects of the world as already interrelated and full of meaning. There is no justification for the assumption that we first experience isolated facts, ... and *then* give them significance" (Dreyfus, 1992, pp. 269–270). Furthermore, Dreyfus emphasizes that "... the situation is organized from the start in terms of human needs and propensities which give the facts meaning, make the facts what they are, so there is never a question of storing and sorting through an enormous list of *meaningless, isolated data*" (Dreyfus, 1992, p. 262). For further elaboration on the basic notion of a "situation", see Dreyfus (1992), pp. 273–282.

[14] This view is implicit in the theories of Marr (1982) and Biederman (1987).

[15] This point is also emphasized by Varela et al. (1991).

features but, primarily, on ways of "knowing-how" and on situated sensorimotor experience.[16] If so, object representations cannot be conceived as merely sensory, but must be envisaged as large-scale sensorimotor patterns.

Neuro-chauvinism: At least among neurobiologists there is a widespread tendency to believe that a complete theory of brain function could fully account for cognitive processes. As Barlow puts it provocatively in his paper on the grandmother cell hypothesis: "Thinking is brought about by neurons, and we should not use phrases like 'unit activity reflects, reveals, or monitors thought processes', because the activities of neurons, quite simply, *are* thought processes" (Barlow, 1972, p. 380). If so, sufficient knowledge of the neuronal correlates of perceptual processes would explain our way of perceiving the world and would cope with the quality of subjective experience. Consequently, it is assumed that neurobiological theorizing provides some kind of privileged access to perception and cognition, and that phenomenological descriptions at the level of the life-world have no right on their own and need to be eliminated from the realm of scientific discourse.[17] Again, the scope of this contribution does not permit us to consider in detail the arguments that have been raised against reductionism. We wish to emphasize, however, that from the point of view we have adopted here this widely accepted dogma needs to be rejected. For several reasons, it appears doubtful that a complete theory of perception can rely on neurobiological descriptions alone. First, there seems to be what could be called the "context-problem": Just looking into brains is not sufficient to specify the contents of mental states, because the latter is always defined only with respect to the environment and *relative to* a situational context in which the respective subject is engaged.[18] Thus, perception cannot by fully explained by an "individualistic" approach referring exclusively to internal states of a single cognitive subject.[19] Second, it has been argued that cognitive processes cannot fully be described at the neural level because exclusive reference to "sub-personal" states and processes means committing a "homunculus fallacy": cognitive acts are executed by *persons* and not by parts of them, and it amounts to a category mistake to describe a brain as "perceiving" or the visual system as "recognizing an object".[20] Third, there is still the unresolved "qualia-problem": just having a description of the neural correlates of pain does not imply knowledge of

---

[16] Our view is supported by the fact that developmental learning of object perception and categorization requires active exploration of the environment and sensorimotor experience, see Held (1965).

[17] In its most extreme version, this view has been made explicit by Churchland, e.g. (1989), pp. 47–66.

[18] We suggest that not only the intentional but also the phenomenal contents of mental states can be subject to such a relational analysis. However, this issue is highly controversial, as reviewed e.g. by Levine (1995).

[19] This point has been elaborated by Burge, see e.g. Burge (1979).

[20] See the chapter "The Homunculus Fallacy" in Kenny (1984), pp. 125–136.

how it feels to be in pain.[21] Thus, it is unclear how the subjective aspects of perceptual experience can be incorporated into an objective neurobiological approach. What these and a number of related arguments suggest is that neurobiological theory cannot deliver the very essence of cognition but just a description of important structural and functional components of cognitive activity. Contrary to the reductionist assumption, we think that the psychological discourse cannot be eliminated in the long run and that, to be valid, a neurobiological theory of perceptual processes must actually be anchored to a phenomenological description of our modes of experience, as they are encountered in the world of daily life.

To summarize, our critical review of the background assumptions of perceptual neurobiology yields the following conclusions: (i) We believe that an adequate theory of perception must account for the constructive aspects of cognition, rather than just describing how environmental information is engraved into connectivity patterns of neural networks. Using Varela's terminology, perception should be understood as the process of "enacting" relevant distinctions in a background without prespecified boundaries. (ii) Accordingly, the perceptual world would not appear as a universe of pre-defined objects but, rather, a field of experience in which objects are individuated according to context, relative to the task and the cognitive agent's concerns. (iii) In this framework, the operation of the visual system is conceptualized as context-dependent selection of relevant information, rather than computation of invariant object descriptions. (iv) Furthermore, representational states are envisaged as large-scale sensorimotor patterns which do not encode fixed stimulus properties, but reflect the capacity of creating perceptual contents. Taken together, we suggest that current neurobiological theories about perception require major modifications. Although the connectionist framework clearly constitutes an advance, its basic assumptions about the nature of perception and cognition still seem to be deficient. Thus, further paradigm shifts will have to occur which enable neurobiologists to relate more appropriately to the actual phenomenology of perceptual processes.

Acknowledgements. We thank Pieter R. Roelfsema and Wolf Singer, with whom the work on the formation of neuronal assemblies has been performed, for many stimulating discussions. We are obliged to Thomas Metzinger for comments on the manuscript and to Raphael Ritz for help with LaTeX style formatting.

---

[21] The famous "qualia argument" has been put forward by Nagel (1974).

References

Barlow, H.B. (1972): "Single Units and Sensation: A Neuron Doctrine for Perceptual Psychology?", *Perception* 1, 371–394

Bechtel, W. and Abrahamsen, A. (1991): *Connectionism and the Mind* (Blackwell, Cambridge MA)

Biederman, I. (1987): "Recognition-by-components: a theory of human image understanding", *Psychological Review* 94, 115–147

Burge, T. (1979): "Individualism and the mental", in French, P.A., Uehling, T.E. and Wettstein, H.K., eds., *Midwest Studies in Philosophy*, Vol. IV (Univ. of Minnesota Press, Minneapolis), pp. 73–121

Churchland, P.M. (1989): *A Neurocomputational Perspective* (MIT Press, Cambridge MA)

Churchland, P.S. and Sejnowksi, T.J. (1992): *The Computational Brain* (MIT Press, Cambridge MA)

DeAngelis, G.C., Ohzawa, I. and Freeman, R.D. (1995): "Receptive-field dynamics in the central visual pathways", *Trends in Neurosciences* 18, 451–458

Dreyfus, H.L. (1992): *What Computers Still Can't Do* (MIT Press, Cambridge MA)

Engel, A.K., König, P., Kreiter, A.K., Schillen, T.B. and Singer, W. (1992): "Temporal coding in the visual cortex: new vistas on integration in the nervous system", *Trends in Neurosciences* 15, 218–226

Felleman, D.J. and Van Essen, D.C. (1991): "Distributed hierarchical processing in the primate cerebral cortex", *Cerebral Cortex* 1, 1–47

Heidegger, M. (1989): *Die Grundprobleme der Phänomenologie* (Klostermann, Frankfurt)

Held, R. (1965): "Plasticity in sensory-motor systems", *Scientific American* 11/65, 84–94

Hubel, D.H. (1988): *Eye, Brain, and Vision* (Freeman, New York)

Kenny, A. (1984): *The Legacy of Wittgenstein* (Basil Blackwell, Oxford)

Knierim, J.J. and Van Essen, D.C. (1992): "Neuronal responses to static texture patterns in area V1 of the alert macaque monkey", *Journal of Neurophysiology* 67, 961–980

Levine, J. (1995): "Qualia: intrinsic, relational or what?" in Metzinger, T., ed., *Conscious Experience* (Schöningh, Paderborn), pp. 277–292

Marr, D. (1982): *Vision* (Freeman, San Francisco)

Merleau-Ponty, M. (1976): *Die Struktur des Verhaltens* (de Gruyter, Berlin)

Nagel, T. (1974): "What is it like to be a bat?" *The Philosophical Review* 83, 435–450

Port, R.F. and van Gelder, T. (1995) *Mind as Motion. Explorations in the Dynamics of Cognition.* (MIT Press, Cambridge MA)

Singer, W. (1995): "Development and plasticity of cortical processing architectures", *Science* 270, 758–764

Singer, W. and Gray, C.M. (1995): "Visual feature integration and the temporal correlation hypothesis", *Annual Review of Neuroscience* 18, 555–586

Spillmann, L. and Werner, J.S., eds. (1990): *Visual Perception. The Neurophysiological Foundations* (Academic Press, San Diego CA)

Ullman, S. (1991): "Tacit assumptions in the computational study of vision", in Gorea, A., ed., *Representations of Vision* (Cambridge University Press, Cambridge), pp. 305–317

Varela, F.J., Thompson, E. and Rosch, E. (1991): *The Embodied Mind. Cognitive Science and Human Experience* (MIT Press, Cambridge MA)

von der Malsburg, C. and Singer, W. (1988): "Principles of Cortical Network Organization", in P. Rakic and W. Singer, eds., *Neurobiology of Neocortex* (Wiley, New York), pp. 69–99

Neural Network Models in Psychology and Psychopathology

Manfred Spitzer

Universitätsklinikum Ulm, Abteilung Psychiatrie III, Leimgrubenweg 12, D-89070 Ulm, Germany. e-mail: manfred.spitzer@medizin.uni-ulm.de

Introduction

From the very inception of the concept of the neuron over a period of almost one hundred years (cf. Breidbach, 1993; Dierig, 1993; Finger, 1994), networks of neurons have been used to capture aspects of higher cognitive functions. Sigmund Freud (1895/1950) and Sigmund Exner (1894) conceived of the flow of energy through networks of neurons in order to explain psychology and psychopathology (see Figs. 1 and 2). Five decades later, McCullough and Pitts (1943/1989) were the first to posit the neuron as an information processing device, and again, suggested psychopathological applications (cf. Spitzer, 1997a). In the 1980s, neural network research gradually invaded almost every field of psychology (cf. Rumelhart & McClelland, 1986), and since the appearance of the landmark paper by Ralph Hoffman in 1987, an increasing number of network models of psychopathology have been proposed. In this paper, applications of neural networks for the understanding of psychology and psychopathology will be discussed. It is argued that the work by psychologists and psychiatrists is not merely a late "add on" to neural network research but rather an integral part of their lines of inquiry. Several examples will be discussed which should demonstrate that neurocomputational models represent the dearly needed bridge between the mind and brain aspect of human nature.

Before the advent of neural network models, conventional digital computers were used for the understanding of higher cognitive functions. However, with its single processing unit, high speed, high reliability, serial processing, and its general architecture, the von Neumann computer (cf. von Neumann, 1958) was very different from any known "biological hardware". This directly implied that higher cognitive functions cannot be achieved by brains in the way digital computers work. It became obvious that many slow and unreliable neurons must work massively in parallel to perform the many computations necessary for even simple word- or face-recognition experiments. Moreover, because of their architecture, computers may fail completely due to a single faulty connection. Brains, in striking contrast, show a more graceful degradation of function when neurons die. For these reasons the idea that computers are very different from biological information processing systems was gradu-

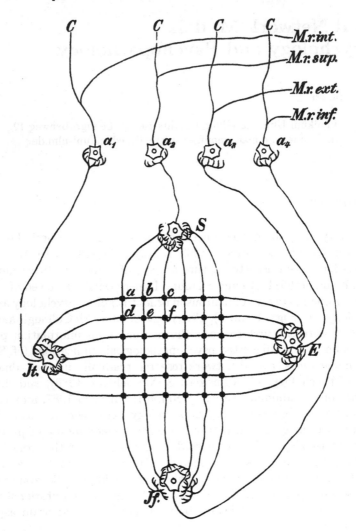

Fig. 1. Neural network by Exner (1894, p. 193), which was meant to explain visual motion detection

ally accepted by psychologists. However, it was not until the 1980s that the idea of neural networks, simulated by using conventional computers in a different way, became the new paradigm for research in cognitive neuroscience. In these networks comprised of idealized neurons as computing devices, each neuron is either active or inactive, and connected to other neurons with *connections* of varying strength. Hence the name *connectionism* for the entire approach to the architecture of ideal systems neurons. Because in these systems neurons work in parallel, and because the work is done at different locations (rather than in only a single CPU), the term *parallel distributed processing*

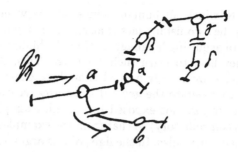

Fig. 2. Neural network drawn by Sigmund Freud in 1895

(PDP) has been used to characterize these models. As David Rumelhart, a major proponent of the connectionist approach, has put it:

> "The basic strategy of the connectionist approach is to take as its fundamental processing unit something close to an abstract neuron. We imagine that computation is carried out through simple interactions among such processing units. Essentially the idea is that these processing elements communicate by sending numbers along the lines that connect the processing elements... The operations in our models then can best be characterized as 'neurally inspired.' ... all the knowledge is in the connections. From conventional programmable computers we are used to thinking of knowledge as being stored in the state of certain units in the system... This is a profound difference between our approach and other more conventional approaches, for it means that almost all knowledge is implicit in the structure of the device that carries out the task rather than explicit in the states of the units themselves" (Rumelhart, 1989, pp. 135–136).

At present, neural network models and simulations of cognitive functions are commonplace in psychology and neuroscience. The models vary a great deal in biological plausibility, complexity, specificity, and explanatory power. Further, they have been used in order to capture features of such different functions as language, attention, memory, and cognitive development. In the next sections, examples from psychology and psychiatry will be discussed.

1 Language Acquisition

The model of language acquisition published in 1986 as part of the two-volume set on PDP models is arguably the most prominent example of PDP models of high-level cognitive functions.

Each of the about 8000 languages on Earth comes with a set of rules, which human beings appear to use routinely in thinking and communicating. Notwithstanding the obvious differences between languages, there are some general principles to which all human beings adhere when speaking. Most noticeably, there is the astonishing fact that children are enormously creative during the process of acquiring language. Although children obviously use the many examples they hear to learn their language skills, the linguist Noam Chomsky (1978) has convincingly argued that these examples are not sufficient for the child to generate the general rules necessarily needed to speak correctly. Children are just not exposed to enough examples (which in any case are sometimes contradictory) so that from the examples alone they could ever generate the necessary rules. In the light of this argument, Chomsky proposed that children must acquire language by some inborn competence, some form of language instinct (cf. Pinker 1994).

Among other things, children must acquire the rules that govern the generation of the past tense from the word stem. Although they do not know these rules explicitly, they can use them creatively. One of these rules, for example, states how to convert the word stem (to sing, to chant) into the past tense (sang, chanted). In the English language, there are two general cases: Many verb stems are converted to the past tense by adding the ending "ed" (to chant – chanted). However, there are exceptions to this rule, and a fair number of verbs have a past tense form that is irregular (sing – sang). While the regular verbs can be changed into the past by the application of a single rule, the irregular forms have to be learnt case by case.

Psycholinguistic studies on the development of language skills in children have demonstrated that children acquire the past tense in certain steps or phases: First they are able to change irregular verbs, which most likely happens by imitation since irregular verbs are also frequent verbs. In a second phase, children appear to acquire the rule which governs the production of the past tense in regular verb cases. However, the children tend to use this rule indiscriminately, i.e., apply it to regular as well as irregular verbs. In this stage, error like "singed" or even "sanged" are common. In this phase, the children are furthermore able to creatively generate the past tense of verbs that do not exist. When asked what the past tense of "to quang" might be, they give the answer "quanged". This capacity has been regarded as crucial evidence that the children must have acquired a rule, which they must have learned to use in the cases they have never heard before. Only in the third phase are the children able to form the past tense of regular as well as irregular verbs, i.e., they have learned the rule and the exceptions to the rule. So they know that it is "to take – took" but "to bake – baked". Once you start to think about how you produce the past tense, you realize how complicated the task really is. Moreover, you realize the enormous task almost all children master in the first few years of their life.

Rumelhart and McClelland (1986) programmed a neural network with 460 input and 460 output nodes, in which every input node was connected to every output node (resulting in $460^2 = 211600$ connections). The input patterns consisted of 420 stems, and the network had to learn the 420 respective

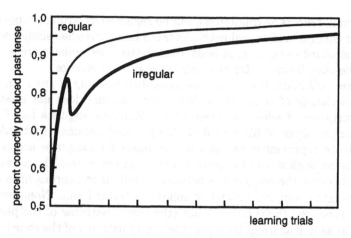

Fig. 3. Result of a neural network simulation of past tense acquisition. The decrease in performance regarding the production of the past tense of irregular verbs, which has been shown to exist in children as well, is clearly visible (redrawn after Rumelhart and McClelland 1986)

past tense forms. Learning was performed by presenting the input layer of the network with a sound-like pattern of the stem, which causes some random activation of the neurons in the output layer. As the desired output (i.e., the sound pattern of the past tense) was known, the actual output and the desired output could be compared. The difference between the actually produced output and the desired output was used to make small changes of the weights of the connections between input and output neurons. By the small adjustment of the weights in the desired direction the network gradually learns to correctly come up with the past tense upon the presentation of the stem.

In fact, after 79900 trials, the network had learnt the correct correspondence between input and output, i.e., had come up with synaptic weights, which led to the correct production of the sound pattern of the past tense upon the presentation of the sound pattern of the stem. Even if new verbs were presented (which corresponds to some extent to the presentation of fantasy verbs to children) the network operated almost flawlessly and produced correctly the past tense of regular verbs in 92% of the cases and even the correct past tense of irregular verbs with an accuracy of 84%. Most notably, the model's learning curve for regular and irregular verbs resembled the respective learning curves of children: The past tense of regular verbs was acquired steadily, i.e., the network produced a steadily decreasing number of errors. In contrast, the past tense of the irregular verbs was first produced increasingly well, until a stage was reached where it actually decreased, only to increase again later.

The fact that the network model, like children, not only gradually improved its performance over time, but also went through similar phases, producing similar errors as children do, can be regarded as strong support for the idea that children and neural networks learn in a somewhat similar way. The striking similarities suggest, at least, that similar mechanisms are at work. If this is the case, however, far reaching consequences emerge.

Notice, for a start, that there was no explicit learning of any rule, but just a gradual change of connections. Moreover, the rule does not exist except as a description of what has been learnt. Whereas linguists like Chomsky and other members of his school of thought had assumed that rules must be explicitly represented *as rules* in the brain for language acquisition to occur, the network model has proved this is not necessary. The network does the task because the connections between hundreds of neurons have changed their strengths, and *not* because it follows a learned rule. Nonetheless, the network produced the correct output (the sound patterns of the past tense) for regular as well as irregular input (the sound patterns of the stem). Regular and irregular verb forms are treated in the very same way by the network. In the network, and by analogy in the heads of speaking human beings, there is neither a rule nor an exception.

Although a number of details of the model proposed by Rumelhard and McClelland have been subjected to criticism (cf. Marcus, 1995; Pinker & Prince, 1988; Plunkett, 1995), the model is very plausible and has been confirmed by further simulation experiments (cf. Hoeffner, 1992). These simulations were able to prove that rule-governed language behavior is possible without any explicit internal representation of the rules.

2   Development and Learning

One of the most fascinating set of ideas has emerged from simulations of interactions between learning and brain development. Usually, computer simulations of neural networks are performed with models that do not undergo architectural changes. They only change in terms of the weights of connections during learning. Only recently have modelers started to study what happens in neural networks which themselves are not static but are subject to developmental change (Elman, 1991, 1994, 1995).

Jeffrey L. Elman turned a standard back propagation network into a so-called recurrent network by adding a context layer (cf. Elman, 1991; Mozer, 1993). This architecture has the effect that the context layer receives as input the activation pattern of the hidden layer without any change. This pattern is then fed back to the hidden layer, *together with the next input pattern*. The hidden layer thereby receives two inputs, new input from the input layer and its own previous state from the context layer.

As the connections from the context layer to the hidden layer are modifiable, patterns can have different effects upon the processing of subsequent patterns. Thus it becomes possible that patterns can have an effect *across time*. This effect is not restricted to one computational step, as the second next input can also be influenced by an input via its effect upon the next one. In fact, such effects across time can be extended over quite a few computational steps. During training, the hidden layer of an Elman network not only forms internal representations of the input patterns which enable the production of the correct output patterns, but it also comes to represent the temporal order of the input patterns.

In an Elman network, the *representation of the context* of an input is therefore possible. A word, for example, is not processed by itself, but instead its processing depends upon the previous words. "Bank" for example, is processed differently in the context of words like "money", "interest", and "Wall Street" as opposed to the context of "river" and "sand". Such effects of context can only be simulated in neural networks if the relevant context is in fact represented within the network, which is the case in neural networks of the Elman type.

Elman trained these networks to learn complex input–output relations, which must be at work whenever complex sentences are understood. As recurrent networks take into account the sequence of patterns, they should be able to process not just single words, but also complex sentences. In brief,

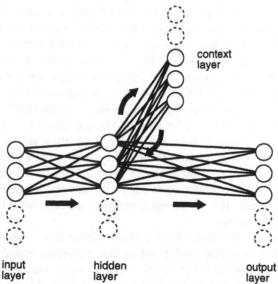

Fig. 4. Architecture of an Elman network. An additional context layer is connected to the hidden layer with one-to-one connections. These connections provide the context layer with a perfect copy of the activation state of the hidden layer. Upon the transmission of the subsequent pattern from the input layer to the hidden layer, this copy of the previous state of the hidden layer is fed back to it via distributed connections whose weights can change. Thus, each patterns influences the processing of the subsequent pattern (arrows indicate the flow of information in the network)

Elman recurrent networks should be able to learn grammar. After training, they ought to, for example, be able to predict the kind of a subsequent word within a sentence. To be able to make such predictions implies knowing grammatic rules. The networks, Elman reasoned, should be able to learn grammar because they take into account sequential (temporal) patterns.

The results of simulations were at first disappointing. When sentences of varying complexity were used as input patterns, the network only came to "understand" simple sentences, in that it was able to make the correct grammatical predictions. Complex sentences, such as sentences with embedded clauses, seemed to be beyond the reach of comprehension by the network. It appeared to be unable to extract the complex rules that structure the sequence of such sentences.

Further simulations, however, proved that the network was able to digest complex sentences, and to learn complex grammar. It did so if *at first simple sentences were used as input*. Once simple sentences had been learned successfully, complex sentences were added to the set of input patterns. Complexity was thereby able to "piggyback" upon simplicity. This may be compared to the way we learn a foreign language by learning simple structures first and then the more complicated ones. We would not get very far by starting out with the most complicated structures in the new and unknown language. Rather, we learn complex structure on top of simple structure.

This result posed a problem when transferred to biological reality. The complex environment of organisms does not come with a "teacher" that provides a carefully controlled sequence of inputs (i.e., learning experiences) such that simple ones are given first, their successful learning is checked and then more complicated ones are introduced. This is quite obvious when children learn their first language. Although studies have shown that caretakers usually engage in some kind of "baby babble" during about the first year of life of the child, most children are for most of their time subject to all kinds of language input. When we talk to children we pay little only attention to whether they receive input that they can process. Hence, if language acquisition in children were dependent upon such carefully orchestrated increasingly complex sets of input, few of us, quite possibly no one, would actually have learned our first language.

It is of interest, therefore, that further simulations demonstrated that there was another condition which led to the successful learning of complex input patterns. If the capacity of the context layer was set to a low level and then increased during learning, the network was able to learn complex grammatical structures. This was the case even if complex input was given to the network from the very beginning of training on. Obviously, the network was again able to extract simple rules of grammar first and then increasingly complicated ones *because it was itself under development from low to high processing capacity* while it learned.

To a network with low context processing power, input sequences governed by complex rules (sentences with embedded clauses) are just like input sequences without any rules. Hence, no rules are learned and the complicated input is processed as if it were nothing but computational noise. To such a network, natural language with its mixture of simple and complex structures, is nothing but a mixture of simple structures and no structure at all. Because of its limited capacity, the network will only extract simple structures from the input, and hence, learn these simple structures. Once these simple structures have been learned and once an increase of the processing power of the context layer has taken place, then it can process somewhat more complicated input structures. It can do so because it is now capable of holding more information together on-line (increased capacity) and also because it can process the more complex input piggyback on the already learned simple structure.

The increased capacity of the system therefore automatically takes care of what a good teacher does, i.e., it provides the system with digestible input. Generally speaking, *a system with increasing capacity during learning can learn complex structures better than a system working at full capacity from the start.*

These results from computer simulations with Elman networks shed new light on what is known about the development of the brain. The extra layer of a recurrent network can be likened to the function of the frontal lobes in the human brain, keeping relevant context on-line. It is therefore important to note that the frontal lobes undergo marked postnatal developmental change in that their connecting fibers become myelinated in the years after birth. As myelination increases the conducting speed of action potentials by a factor of about 30, the neurons in the frontal lobe can be conceived as existent at birth, but only incrementally going on-line for processing. As Elman's simulations show, in contrast to a fully developed brain at birth, a postnatally developing brain allows for the learning of more complex structures. The brain of a newborn baby is highly immature, and the process of extra-uterine maturation affects mainly the frontal cortex, i.e., the area in which the highest and most complex mental functions, including language, are localized. In addition, the frontal cortex is the part of the brain which is connected to other parts of the brain analogous to an extra feed back loop, as the context layer in an Elman network is connected to the hidden layer. The frontal lobes are the site of working memory, where immediately relevant information is kept on-line to bear upon what happens in the here and now. This part of the brain comes on-line in children during nursery school and school.

With respect to language acquisition, this implies that the not fully developed frontal lobes of the newborn baby are *not an obstacle, but rather a computational prerequisite* for learning the complexities of language. We may add, that this may be the case for any other complex cognitive function as well.

3 Hopfield Networks and Hallucinations

The use of neural networks for the study of psychopathological phenomena is not new, but rather has a rich historical tradition. One striking feature of this tradition is that from the very inception of the idea of the neuron, psychiatrists have used the notion of networks and their pathology to account for psychopathological phenomena. Moreover, many advances in neural network research were either made by psychiatrists or were put forward in relation to psychopathology. In other words, neural network studies of psychopathological phenomena are by no means a "late add on" to the mainstream of neural network research, but rather have always been at the heart of the matter.

At present, there is hardly a psychopathological symptom or disorder for which no network models exist (cf. Reggia et al., 1997). This is surprising, since the field of neural network computer modeling of psychopathology is only about 10 years old. In 1987, the Yale psychiatrist Ralph Hoffman published the first major paper on neural network models of psychopathology. While he has refined his models in subsequent papers (cf. Hoffman et al., 1995), his early work was a landmark in that it laid out how simulations and psychopathology could be linked.

Hoffman trained Hopfield networks to store input patterns. In these networks every neuron is connected to every other neuron except itself. These connections make the network behave such that upon the presentation of an input it will settle into a certain state, which can be interpreted as the stored memory trace. Such networks, however, have a limited capacity.

The way in which recurrent networks perform can be characterized by saying that upon the presentation of a given input, the pattern of activation of the neurons in the network, converges to a specific output state, a stable activation pattern of the neurons in the network called attractors. If the sequence of states upon activation by an input is closely scrutinized, one observes that the pattern of activation changes such that it becomes increasingly like the stable state that most closely resembles the input pattern. The network converges to the attractor that is closest to the input pattern. All possible states of the network can be metaphorically likened to a landscape of energy, where the attractors form valleys between mountains of non-stable network states.

According to the simulations performed by Hoffman, information overload of the networks leads to network behavior that can be likened to hallucinations in schizophrenic patients, in that the spontaneous activation of some stored patterns occurs. Hoffman describes how, when real biological neural networks are diminished in size and/or storage capacity by some pathological process, they can no longer handle the experiences of the person. Eventually, information overload will lead to the deformation of the landscape, described by Hoffman as follows:

"Memory overload [...] causes distortions of energy contours of the system so that gestalts no longer have a one-to-one correspondence with distinct, well delineated energy minima" (Hoffman, 1987, p. 180).

In addition to the deformation of the structure of the network, new attractors are formed, which Hoffman called "parasitic". Such parasitic attractors are produced by the amalgamation of many normal attractors, whereby particularly stable attractors, i.e., "deep valleys" in the energy landscape, are produced. These deep valleys are therefore the end state of the system starting out from a large number of positions. In this way, these parasitic attractors are reached with almost any input to the system. According to Hoffman, these attractors are the basis of voices and other major symptoms of schizophrenia:

"These parasitic states are the information processing equivalents of 'black holes in space' and can markedly disrupt the integrity of neural network functioning by inducing 'alien forces' that distort and control the flow of mentation. If a similar reorganization of the schizophrenic's memory capabilities has taken place, it would not be at all surprising for him to report one or more schneiderian symptoms, namely, that he no longer has control of his thoughts, or that his mind is possessed by alien forces" (Hoffman 1987, p. 180).

Hoffman and coworkers (1995) have proposed other more detailed models of schizophrenic hallucinations which encompass the specific effects of neuronal loss in the frontal lobes and the effects of dopamine. Nonetheles, the model just discussed led to the development of an entire new field which may be called *neurocomputational psychopathology*.

4   The Neuropathology of Alzheimer's Disease

In this section, the work of Michael Hasselmo from Harvard University on the interface of neurophysiology and artificial neural network dysfunction is discussed. They exemplify how detailed network models can be used to understand patterns of neuropathology.

The neuropathologist Alois Alzheimer first described changes in the brain of a patient who had suffered from progredient severe dementia. Similar microscopic changes were subsequently found in the brains of demented elderly patients. At the beginning of such dementia processes, the memory for new facts is particularly impaired whereas other mental functions may be fine. Later in the course of dementia, long-term memory is also impaired and during the final stage, the person becomes completely incapacitated, is no longer able to recognize their closest relatives, and requires permanent physical care.

Hasselmo starts with the fact that in neural network simulations the learning phases have often to be distinguished from the retrieval phase. During learning, the spreading of activation through the network must be prevented to some degree, otherwise synaptic activation and change would spread like an avalanche through the entire network. Hasselmo was able to show that, whenever overlapping patterns have to be learned, which is what happens in biological systems, such *runaway synaptic modification,* as he called the phenomenon, occurs (see Fig. 5). The end result of the process is that every input pattern activates all output neurons. This is equivalent to a network that has not learned anything, since learning always implies discrimination. A network that leads to the activation of all output neurons in response to an input is computationally useless.

In most computer simulated network models, runaway synaptic modification is prevented by inhibiting the spread of activation within the network during learning, since such spread interferes with learning, as we have just seen. Only during the process of retrieval (i.e., whenever the network carries out what it has previously learned) does the spread of activation through it become essential. This change in the spread of activation within computer simulations of networks can be implemented in various ways by mathematical procedures. The question, however, is how runaway synaptic modification is prevented in biological systems such as the human brain.

To repeat the problem, during learning, the network must be sensitive to activation from outside, but the spread of activation via internal connections must also be suppressed, while during recall, activation must be able to spread through the network via internal connections. Within this framework, recent neurobiological findings on the role of acetylcholine in the dynamics of cortical activation are of special importance. It has been known for a long time that *acetylcholine* is related to processes of learning and memory. The substance is produced by a small number of neurons which are clustered together in the

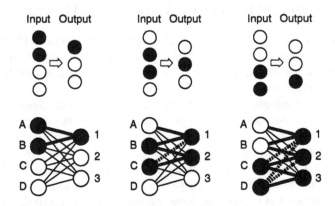

Fig. 5. (Top) Relation between input and output patterns to be learned by the network (active neurons are drawn in black, inactive neurons in white). (Bottom) Two layer network learning the input-output function depicted at the top

nucleus basalis Meynert. From there, acetylcholine can be spread out through almost the entire brain via tiny fibers that proceed to the cortex.

Experiments in slices of the entorhinal cortex have demonstrated that acetylcholine selectively suppresses excitatory synaptic connections between neurons of the same cortical region. In contrast, signals from other cortical areas can pass through synapses unimpaired. Thus, acetylcholine has the very function needed in neuronal networks to prevent runaway synaptic modification such that learning of new and overlapping patterns can occur. In order to do its work, acetylcholine must be liberated during learning, but not retrieval. This in turn presupposes the existence of a fast acting mechanism that evaluates and detects the novelty of an incoming stimulus pattern and sends the result of this evaluation to the nucleus basalis Meynert. The existence of such a mechanism in the brain is conceivable and has in fact been assumed for other reasons as well, even though there is no detailed view on it thus far (cf. Hasselmo, 1994, p. 22).

For many years it has been known that in Alzheimer's disease the amount of acetylcholine in the brain is decreased. However, before Hasselmo's model it was not clear how the lack of this substance has an effect on learning. Within the framework of Hasselmo's model, however, one can deduce what happens if acetylcholine is lacking in the brain: The model explains why in Alzheimer's disease the learning of new information is affected severely at an early stage. Furthermore, it is known that overly active neurons using the transmitter glutamate may be damaged by their own excessive activity. This phenomenon is referred to as *excitotoxicity* and is well described for glutamatergic neurons. For years, such excitotoxicity has been implicated in the neuronal loss observed in Alzheimer's disease. The functional role of acetylcholine in the prevention of runaway synaptic modification explains how the lack the substances can lead to excitotoxicity and neuronal loss.

If acetylcholine is decreased or lacking, the number of associations that form within neural networks increases. This causes the increased activation of neurons within the network upon the presence of any input. In short, acetylcholine puts the brakes on cortical excitation during encoding, and if this process is malfunctioning, cortical excitation increases. Hence, any new learning causes changes in both desired and undesired synaptic connections, which not only interferes with learning, but also leads to the increased activation of glutamatergic neurons. As just mentioned, this activity can be toxic for the neurons themselves.

Hasselmo's model predicts the spatial pattern of neuropathology observed in the brains of deceased Alzheimer's patients. Such pathology should occur in those brain areas that are highly involved in associative learning like the hippocampus. The typical pathological changes of Alzheimer's disease, most importantly the Alzheimer tangles, become visible earliest and most frequently in the hippocampus. Even within the hippocampus, the Alzheimer related pathology occurs in regions which are known to have the most easily modifiable synaptic connections (notably those in which the process of long term potentiation was first discovered). Hasselmo (1994) discusses a number of further neuropathological changes which can be explained easily within his model, which, like other neural network models, put together a number of otherwise inexplicable or unrelated characteristics of the disorder.

5 Concluding Remarks

An increasing number of studies now demonstrate how higher cognitive functions and their respective pathology can be simulated with neural networks. These studies have sometimes produced astonishing results. Obviously, the fact that a certain task can be simulated on a computer running a neural network model does not prove that the task is implemented in the brain as it is in the network (cf., Crick 1988, 1989). In contrast to explanatory hypotheses and theories, a simulation has the advantage that it can be subjected to experimental manipulations and empirical tests of the effects of these manipulations.

In other words, a theoretical explanation can only be confronted with new data which are predicted more or less well by the theory. A simulation model, however, allows for the introduction of changes, and for the actual simulation of what results from these changes. In "playing" with the model, one can generate predictions about the real world, and check these predictions by confronting them with new observations and experiments. In short, the model allows for *experiments with the mind*. For example, parameters of the network, such as the patterns of connection, the activation function of the neurons, or its size, can be manipulated just as well as the input signals, and the resulting behavior can be studied in detail. Such experiments are either impossible or very hard to carry out on real neuronal systems. A network can even be damaged to various degrees (neurons and connections can be set to zero), and the effects of such damage upon performance can be tested under different circumstances, for example, different levels of informational load. We have already discussed an example, in which the learning history, i.e., the changes in performance over time, is highly informative, especially if there are specific differences which depend upon input characteristics. The more detailed the predictions are and the more sophisticated the model, the more informative are the results of such simulations. Of particular interest are counterintuitive results, which simulations may produce and which help to direct the attention of the researcher to phenomena and functional relations which would otherwise have escaped attention completely. In short, the possibilities of network simulations are endless. They are only limited by the fantasy and creativity of the experimenter.

Whether a network simulation model behaves like a nervous system proves nothing about how the real nervous system works. In particular, it cannot be taken as proof that the real nervous system functions in the same way as the simulations. With this caveat, a network model can nonetheless demonstrate operating principles that might be at work in real nervous systems, and in fact, such a model may be the only way to detect these principles. If a model simulation behaves strikingly similarly to a real nervous system upon the introduction of damage, for example, and if the model generates unexpected predictions about the behavior of the biological system, which are found to be correct upon subsequent careful examination, one can hardly escape the compelling plausibility of such a model.

References

Breidbach, O. (1993): Nervenzellen oder Nervennetze? Zur Entstehung des Neuro-nenkonzepts, in: E. Florey and O. Breidbach, (eds.): *Das Gehirn – Organ der Seele?* (Akademie-Verlag, Berlin), pp. 81–126

Chomsky, N. (1978): *Rules and Representations* (Columbia University Press, New York)

Crick, F. (1988): *What mad pursuit* (Basic books, New York)

Crick, F. (1989): The recent excitement about neural networks. *Nature 337*, 129–132

Dierig, S. (1993): Rudolf Virchow und das Nervensystem. Zur Begründung der zellulären Neurobiologie, in: E. Florey and O. Breidbach, (eds.): *Das Gehirn. – Organ der Seele?* (Akademie-Verlag, Berlin), pp. 55–80

Elman, J.L. (1990): Finding structure in time, *Cognitive Science 14*, 179–211

Elman, J.L. (1991): Incremental learning, or The importance of starting small, in: *Proceedings of the Thirteenth Annual Conference of the Cognitive Science Society*, p. 443–448 (Erlbaum, Hillsdale, N.J.)

Elman, J.L. (1994): Implicit learning in neural networks: The importance of starting small, in: C. Umilty and M. Moscovitch (Eds.) *Attention and Performance VI Conscious and nonconscious information processing* (MIT Press, Cambridge, MA) p. 861–888

Exner. S (1894): *Entwurf zu einer physiologischen Erklärung der psychischen Erscheinungen*, Deuticke, Leipzig, Wien

Finger, S. (1994): *Origins of neuroscience. A history of explorations into brain function* (Oxford University Press, Oxford)

Freud, S. (1978): Project for a scientific psychology (1895). *The Standard Edition of the Complete Psychological Works of Sigmund Freud*, vol. 1 (Hogarth Press, London) pp. 283–397

Hasselmo, M.E. (1994): Runaway synaptic modification in models of cortex: Implications for Alzheimer's disease. *Neural Networks 7(1)*, 13–40

Hobson, J.A. and McCarley, R.W. (1977): The brain as a dream state generator, *American Journal of Psychiatry 134:* 1335–1348

Hoeffner, J. (1992): Are rules a thing of the past? The acquisition of verbal morphology by an attractor network, *Proceedings of the Fourteenth Annual Conference of the Cognitive Science Society* (Erlbaum, Hillsdale, New Jersey) p. 861–866

Hoffman, R.E. (1987): Computer simulations of neural information processing and the schizophrenia-mania dichotomy, *Archives of General Psychiatry 44*, 178–185

Hoffman, R.E., Rapaport. J., Ameli, R., McGlashan, T.H., Harcherik, D. and Servan-Schreiber, D. (1995): The pathophysiology of hallucinated 'voices' and associated speech perception impairments in psychotic patients, *Journal of Cognitive Neuroscience* 7 479–496

Marcus, G.F. (1995): The acquisition of the English past tense in children and multi-layered connectionist networks, *Cognition 56,* 271–279

McCullouch, W.S. and Pitts, W. (1943/1988): A logical calculus of the ideas immanent in nervous activity, in: J.A. Anderson and E. Rosenfeld (eds.): *Neurocomputing. Foundations of Research* (MIT Press, Cambridge, MA) pp. 18–27

Mozer, M.C. (1993): Neural net architectures for temporal sequence processing, in: A. Weigand and N. Gershenfeld (Eds.) Predicting the future and understanding the past (Addison-Wesley, Reading, MA)

Neumann, J.V. (1958): *The computer and the brain* (Yale University Press, New Haven)

Pinker, S. (1994): *The language instinct. How the mind creates language* (William Morrow and Company, New York)

Pinker S. and Prince A. (1988): On language and connectionism: An analysis of a parallel distributed processing model of language acquisition, *Cognition 28* 73–193

Plunkett, K. (1995): Connectionist approaches to language acquisition, in: Fletcher, P. and MacWhinney, B. (Eds.): *Handbook of Child Language* (Blackwell, Oxford) pp. 36–72

Plunkett, K. and Marchman, V. (1991): U-shaped learning and frequency effects in a multi-layered perceptron: Implications for child language acquisition, *Cognition 38,* 1–60

Reggia, J.A., Ruppin, E. and Berndt, R.S. (1997): *Neural Modeling of Brain and Cognitive Disorders* (World Scientific, Singapore)

Rumelhart, D.E. (1989): The architecture of mind: A connectionist approach, in: Posner MI (ed.): *Foundations of cognitive science* (MIT Press, Cambridge, MA) pp. 133–159

Rumelhart, D. and McClelland, J.L. (1986): On Learning the past tense of English verbs, in: D.E., Rumelhart and J.L. McClelland, PDP Research Group: *Parallel distributed processing: Explorations in the microstructure of cognition, Vol. II* (MIT Press, Cambridge) pp. 216–271

Spitzer, M. (1997): The History of Neural Network Research in Psychopathology, in: D. Stein (ed.) *Neural Networks and Psychopathology* (Cambridge University Press, Cambridge, UK) (in press)

Questions Concerning Learning in Neural Networks

Ion-Olimpiu Stamatescu

FEST, Schmeilweg 5, D-69118 Heidelberg, Germany,
and
Institut f. Theoretische Physik, University of Heidelberg,
D-69120 Heidelberg, Germany.
e-mail: stamates@thphys.uni-heidelberg.de

1   Introduction

Learning has always been related to the development of intelligence, but there are different views about the dynamics of this relation. A conservative view of learning, developed in the frame of cognitive psychology is that of "knowledge acquisition". In a traditional behaviorist approach, on the other hand, learning is related to "adaptation", "homeostasis", etc. Finally, the modern developments in neural science reveal the very complex behavioral, cognitive and neurophysiological structure of some basic mechanisms of learning which are active already at the level of conditioning and other "simple" cognitive activities.

The "neural science" view distinguishes two types of learning and correspondingly two types of memory: implicit and explicit. Implicit learning is typically involved in acquiring skills, but it also takes part in abstract, cognitive processes, as long as it proceeds without making explicit the learning tasks and steps. So, for instance, amnesic patients perform worse than normal subjects on the task of recalling words (learned a while before) when both are cued with the first 3 letters of the word. They perform, however, normally when asked to freely associate words with the 3-letter cues, which indicates that the learning has biased the hidden associative structures in the normal way (Kandel et al., 1995). Implicit learning can be nonassociative or associative and then typically proceeds by reinforcement.

Explicit learning, like explicit memory is also rather complicated. It is an active process which transforms the original information both during storage and during recall by reordering, condensation, reconstruction, and, in turn, by comparison, suppositions, associations – which suggests that all conscious cognitive activity is basically conducted by the assumption of/search for coherence. This aspect reappears at each level of our activity, it is "ein regulatives Prinzip unseres Denkens" ("a regulative principle of our thinking"), "es spricht [das Vertrauen in die Gesetzmäßigkeit,] in die vollkommene Begreifbarkeit der Welt aus. Das Begreifen [...] ist die Methode, mittels deren unser

Denken die Welt sich unterwirft, die Tatsachen ordnet ..." states Hermann
von Helmholtz in his seminal lecture "Die Tatsachen in die Wahrnehmung"
(1878)[1]. We also see in the above aspect how intimately connected learning
is to all cognitive activity.

One can also distinguish between a short- and a long-term memory, which
involve not only different cognitive steps but also different brain pathways.
Again, long-term potentiation (of synapses) in associative and non-associative
combinations[2] appears to be the neural basis of the corresponding type of
learning. Explicit and implicit learning interact strongly and some explicit
memory can be transformed into implicit memory by usage.

We should, however, stress that the biology of learning, besides providing
an impressive amount of knowledge, has also raised very many new questions,
and that many of the hypothezised mechanisms are still speculative.

2   Learning in Neural Network Models

Since neural networks (NNs) are repeatedly discussed in this book (cf. the
chapters by H. Horner and R. Kühn, W. Menzel, A. Engel and P. König,
and M. Spitzer, cf. also the Introduction) we shall restrict ourselves here to
recalling some principal features that are relevant for our discussion.

From the point of view of cognition, neural networks correspond to a new
paradigm, called "connectionism". In contradistinction with the "physical-
symbol-system" paradigm (corresponding to the "standard AI"), which is
a top-down approach, "connectionism" represents a bottom-up approach in
which "thinking" *results* from the activity of big assemblies of highly interconnected units. Assuming them to incorporate background mechanisms for
cognition (Rumelhard and McClelland 1986) NNs should be adequate for
modeling various problems, such as associative recognition of visual patterns
or semiotic processes, e.g., learning languages. The interesting features of
neural networks are: associativity, stability under small disturbances, easy
adaptation to new requirements, "self-programming" as a result of learning.
Of course, neural networks models also have a number of shortcomings related, e.g., to the statistical character of their performance in applications
in expert systems, for instance. [3] An especially weak point seems to be the
difficulty in explicitly incorporating high level cognitive procedures or in de-

---

[1] "it expresses the faith in the full understandability of the world. Understanding
is the method by which our reason conquers the world, orders the facts ...", see
Helmholtz (1987) (ad hoc translation).

[2] LTP – long-term potentiation – increases the efficacity of a synapse in producing
a post-synaptic potential under strong and/or repeated pre-synaptic (action)
potential, either in association with activation of the post-synaptic neuron or
independently of it.

[3] A way to control these uncertainties by turning them into exact confidence limits
for the results is described in Berg and Stamatescu (1998).

scribing their achievements in terms of strategies, etc. (this is a typical problem for a bottom-up approach, just as context dependence, associativity, etc. was a problem for the top-down approach of standard AI).

Detailed information about the brain and its activity is provided by modern neurophysiology (see, e.g., Kandel et al., 1995). The brain is a highly interconnected system (for the about 10^{11} neurons there are between a few and as many as about 10^5 synapses per neuron – one quotes an average synapse number of 10^4). A typical feature of the brain is the diversity of mechanisms which seem to play a similar role and the variability in the implementation of one or another function: hundreds of types of ion channels which control the synapse efficiency, various kinds of synapse plasticity, various kinds of organization, etc. – together with some invariant main characteristics like the predominantly threshold behavior of neurons and the transmission of information via synapses.

The principal aspect of learning in NNs is the modification of the synapses according to one of two types of algorithms:

- *supervised learning* makes the change depending on the reaction of the network (namely, such that its functioning approaches the desired behavior), and
- *unsupervised learning* makes the change depending solely on the coincidence of the firing state of the pre- and post-synaptic neuron in the desired input–output situation.

Both procedures can be related to the biologically motivated *Hebb rule* which asserts that there is a feedback from a firing neuron onto all the synapses which contributed positive presynaptic potentials to its firing and this feedback increases the corresponding weights (makes the synapses more efficient). A formal implementation of this rule, e.g. using the $\sigma = \pm 1$ (*Ising*) representation for two-state neurons gives for the change of the synapse between pre-synaptic neuron j and post-synaptic neuron i:

$$C_{ij} \rightarrow C_{ij} + \epsilon\, \sigma_i \sigma_j \,, \tag{1}$$

where C_{ij} are the weights (synapses) and ϵ a parameter describing the speed of the adaptation of the weights.

If one considers a fully connected network σ_i, $i = 1, \ldots, N$ where a pattern ξ_i is repeatedly presented on the net the weight changes are described by eq. (1) with σ replaced by ξ. Training the net with a set of patterns ξ_i^μ, $\mu = 1, 2, \ldots, N_p$ is therefore represented by:

$$C_{ij} \rightarrow C_{ij} + \epsilon \sum_{\mu=1}^{N_p} \xi_i^\mu \xi_j^\nu \,, \tag{2}$$

which is the basic unsupervised learning algorithm for such a network. In the supervised learning case the application of the above rule is made dependent on the actual "output" σ_i produced after presentation of each "input" ξ_i^μ.

The neurophysiological mechanism of "potentiation" mentioned before could be thought of as being the basis of the Hebb rule, but on the one hand the neurophysiology of the synapse plasticity is very complex and on the other hand the simple rule (1) must usually be further refined to provide good learning properties (the parameter ϵ and the way of performing the $\sum_\mu$ in (2) are typically matters of tuning).

In connection with their application to AI problems (expert systems, robots, pattern recognition) powerful learning algorithms for artificial neural networks have been developed which are not motivated by biological similarity. For multi-layer perceptrons, "back-propagation"-type algorithms can be very efficient (see, e.g., Rumelhart, Hinton and Williams in Rumelhart and McClelland 1986, Wasserman 1993, etc.). More refined networks and learning rules include "sparse coding", "resilient propagation" or even "evolutionary algorithms" governing the development of the architecture itself (see, e.g., the chapters of Horner and Kühn, and of Menzel). Finally, one can also consider the problem of "unspecific reinforcement", where the evaluation gives only the average error for a whole set of training steps (see next section).

Unsupervised learning, in particular, appears biologically supported and moreover has the appeal of "self-organization" (e.g., in the *Kohonen networks*, cf. Kohonen, 1989) which is probably a very important feature of the brain. Nevertheless, supervised learning can also be active biologically – and not only in the selective sense that unsuccessful learning may lead to the disappearence of the organism. Present neurophysiological information does not exclude the possibility of connections influencing the modifiability of synapses in accordance to experiences of the organism (pain, satisfaction, etc). In any case, one thing to be noticed is that the biological basis of the Hebb rule is very complicated. We find a number of different biochemical mechanisms contributing in the synaptic modulation and the latter can be of many types: selective and non-selective, associative and non-associative, short term, medium term and long term, etc. Neurons in different regions of the brain show one or another feature predominantly, sometimes in groups (bundles of associative pathways followed by bundles of non-associative pathways) or intermixed.

Learning in artificial intelligence systems may incorporate characteristics both of natural evolution and of natural learning. We can therefore put the problem of modeling learning into a rather general setting. We shall consider here some examples.

As is well known (see, e.g. the chapter of Horner and Kühn) if the ratio between the number of stored patterns and the average number of synapses per neuron (N_p/N for a fully connected Hopfield network) becomes larger than about 0.14, the network enters the so-called "spin glass" phase where retrieval of learned patterns is no longer possible. However, in normal situations living systems are literally overwhelmed with patterns. In fact the reaction of a network to such a situation is very simple and in a sense "natural": It retains the common features, that is, what is "persistent" in the information it receives and "ignores" the random stimuli.

Another aspect is that of learning from complex experiences, where the feedback is not immediate and differentiated: Living organisms often only experience the final success of a long series of acts, or the average success of different collaborating procedures. They have to find out by themselves how good the *separate* steps or the particular procedures were, in order to improve on their performance. A perceptron learning algorithm based on the Hebb rule but with unspecific assessment of the error (only the average error for each group of trials is provided) indeed succeeds in adapting the network so as to achieve high performance – see Stamatescu (1996). On the other hand, more complicated problems may appear difficult to implement, e.g., that of improving the behavior in complex situations on the basis of a crude "worse/better" assessment of the final achievements of long series of actions. While this may be straightforward in standard AI – in Mlodinow and Stamatescu (1985), for instance, such a model was developed for a "robot" navigating on a table with obstacles – it appears much more difficult for NN modeling since it is not easy to identify and codify behavioral features. In fact this seems to be a typical case for "marrying" neural networks with classical AI procedures (Riedmiller et al 1997).

Finally we can consider the possibility of using neural networks to understand more refined features of learning seen as a natural behavior – for instance, the role of certain architectures in connection with certain learning problems – see, e.g., McClelland et al (1995) and further examples in the chapter by Spitzer.

3 A Model for Learning

We have seen that various aspects of "natural" learning find a direct representation in NN learning. We can reverse the question and ask what perspective on learning is provided by the NN models.

A well known statement about learning in AI is that "one learns only what one nearly knows", with "nearly" being sometimes understood as only a reordering. This is an important statement, both practically and theoretically, and the creative processes associated with the emergence of a new hypothesis are often based on the reassessment of known facts. But we should not forget that a very essential character of learning resides in the appropriation of something fundamentally alien.[4] Combinatorics and reassessment are essential in "understanding". But there is an even more elementary level there, the appropriation of something genuinely *new*, without which we cannot speak of learning. It may be interesting to note that in very different frames of thinking neither, say, biologists nor theologians appear to have any difficulty

---

[4] This is *not* in contradiction with the "small-step principle" itself.

in considering the appropriation of something alien in biological processes, or in social cross-cultural encounters, respectively[5].

Although these observations are not new, it may be useful to stress the perspective they introduce by specifying a learning model with at least two classes of procedures which can be understood as two stages[6]:

A. *Assimilation:*

A1. incorporation – simple taking over of alien elements without any assessment of their functions;

A2. simulation – reproduction of "strange" behavior without previous evaluation of its effects;

A3. production – of experiences by blind acting on the environment;

and:

B. *Understanding:*

B1. decomposition – of the newly acquired experiences into functional subparts, recognition of already known structures, interpretation in terms of the known and acceptance of the unknown;

B2. reconstruction – of functional structures taking account of the new elements;

B3. creation – imagining and proposing hypotheses.

The class B procedures can be directly related to the more standard approach of knowlegde acquisition while those of class A may be less explicit in this context. On the other hand, it is just the first step (A1) which is so naturally realized in the Hebb rule for learning in neural networks that it may appear trivial. This is, however, an illusion: the Hebb rule is both nontrivial and decisive for the success of life, as it is indicated by the fact that its assumed biochemical basis (associative long term potentiation) is highly complicated and that it has been steadily developed throughout evolution. The second step (A2) is, of course, implicit in the activity of the network. Steps A3 and B3 explicitly introduce an active element and are very important since they are the principal carriers of the creative component in learning – they are mentioned here especially to illustrate how this creative element shows up at the two stages. As universal features of learning behavior the steps of class A can, of course, also be realized in classical AI, but the NN implementation is very natural. However, class B procedures are not easily made explicit in neural networks learning.

---

[5] See, e.g. Geraci (1997), Moltmann (1992).

[6] Other perspectives can also be considered of course, e.g. *adaptation*. For the discussion intended here, however, we can restrict ourselves to these two stages. Their names are intended to emphasize the conceptual difference.

4   Learning and Neural Reductionism

As discussed in the introduction to this book the direct reduction "mental states–brain states" implied by an attempt to "explain" the abstract principles of thinking with help of neurophysiology appears incomplete since even the definition of "mental states" assumes a cultural frame of interactions – see also Putnam (1992), compare with Rumelhart and McClelland (1986), Churchland (1989); see also the the chapter by A. Engel and P. König. We suggested that the similarity between features of the proceedings at the neural activity level (logic functions – in the sense of McCulloch and Pitts 1943 –, associativity, generalization, fuzziness, etc.) and corresponding features of the thinking activity at the mental level can only be viewed as a compatibility basis on which a non-linear process acts building up a hierarchy of (meta-)structures. In this process learning plays an essential role. It is interesting therefore to develop this discussion a bit further in connection with the question of learning.

Here again a "transparency" between levels can be observed, a similarity between learning processes at various levels. In this sense the NN approach to learning supports the view that the dividing line between "brain processes" and "mental processes" becomes relative. In the fundamental mechanisms of associative/nonassociative potentiation, the basis for the cognitive functions acting at various levels of explicit or implicit learning – the Hebb rule and related algorithms – is already found at the most elementary, cellular level.

From the point of view of the model suggested in the previous section the neural science approach seems mainly relevant for the first stage (class), called there "learning by assimilation" (while the second stage, "understanding", involves symbol manipulation). In connection with "assimilation", a problem seems to arise, namely, how to ensure an appropriation of alien elements, and one's own corresponding evolution, while preserving one's own "identity" as individual. Individuality is needed in order to speak of an organism, of a species, etc. The question about "individuality" is, of course, a rather difficult one. Now, philosophical considerations tend to situate it in a force field between internal and external determinations, as for instance the opposition between unity (in oneself) and reference (to others), or that between existence (unstructured) and difference (complex) – see, e.g., Boehm and Rudolph (1994). It should be understood as a dynamical question and it should involve continuity and development. In the context of our discussion, such a dynamical concept of individuality should be ensured by the interplay between the various steps, in particular between the two stages. This emphasizes the collaboration between them understood as independent procedures.

Again we observe the same peculiar character of the "reduction" mentioned in the introduction: On the one hand neuroscience describes the structure which provides the "microstructure" at all higher levels of cognitive activity, specifically, learning – and thus ensures a certain kind of compatibility between these levels. And on the other hand the activity at these higher levels cannot be separated from the intersubjective context, which makes them irreducible.

References

Berg, B.A. and Stamatescu, I.-O. (1998): "Neural Networks and Confidence Limit Estimates" in Meyer-Ortmanns, H., and Klümper, A. eds, (1998): *Field Theoretical Tools for Polymer and Particle Physics* (Springer, Heidelberg)

Boehm, G. and Rudolph, E. (eds) (1994), *Individuum*, (Klett-Cotta, Stuttgart)

Cassirer, E. (1987): *Zur Modernen Physik*, (Wissenschaftliche Buchgesellschaft, Darmstadt)

Churchland, P.M. (1989) *A Neurocomputational Perspective*, (MIT Press, Cambridge, MA)

Geraci, G. (1997), "A molecular mechanism for the evolution of eukariotes", in Proceedings of the Conference *Philosophy of Biology* of the *Académie Internationale de Philosophie des Sciences*, Vigo, 1996

Helmholtz, H. v. (1987): *Abhandlungen zur Philosophie und Geometrie* (Junghans-Verlag, Cuxhaven)

Kandel, E.K., Schwartz, J.H., Jessel, T.M. (1995): *Essentials of Neuroscience and Behavior* (Appleton and Lange, Norwalk)

Kohonen, T. (1989): *Selforganization and Associative Memory* (Springer, Berlin)

McClelland, J.L., McNaughton, B.L., O'Reilly, R.C. (1995): "Why are there complementary learning systems in the hippocampus and neocortex: insights from the success and failures of connectionist models of learning and memory", *Psychological Review, 102*, 419

McCulloch, W.S. and Pitts, W.A. (1943): "A Logical Calculus of the Ideas Immanent in the Nervous Activity", *Bull. Math. Biophys. 5*, 115.

Mlodinow, L. and Stamatescu, I.-O. (1985): "An evolutionary procedure for machine learning", *Int. Journal of Computer and Inform. Sciences, 14*, 201

Moltmann, J. (1992): "Die Entdeckung der anderen" (The discovery of the others), in Jürgen Audretsch (ed), *Die andere Hälfte der Wahrheit* (C.H. Beck, München).

Putnam, H. (1992): *Representation and Reality* (MIT Press, Cambridge, MA)

Riedmiller, M., Spott, M., Weisbrod, J. (1997): "First results on the application of the Fynesse control architecture" IEEE 1997 Int. Aerospace Conference, Aspen; "Fynesse: A hybrid architecture for self-learning control" Techn. report. University of Karlsruhe. Submitted for publication.

Rumelhart, D.E. and McClelland, J.L. (eds) (1986): *Parallel Distributed Processing: Explorations in the Microstructure of Cognition* (MIT Press, Cambridge, MA)

Stamatescu, I.-O. (1997): "The Neural Network Approach" in Proceedings of the Conference *Philosophy of Biology* of the *Académie Internationale de Philosophie des Sciences*, Vigo, 1996

Wasserman, Ph.D. (1993): *Advanced Methods in Neural Computing* (Van Nostrand Reinhold)